Social Science Data Analysis

Florian G. Hartmann · Johannes Kopp ·
Daniel Lois

Social Science Data
Analysis

An Introduction

 Springer

Florian G. Hartmann
Paris Lodron University Salzburg
Salzburg, Austria

Johannes Kopp
Universität Trier
Trier, Germany

Daniel Lois
University of the Bundeswehr Munich
Neubiberg, Germany

ISBN 978-3-658-41229-6   ISBN 978-3-658-41230-2   (eBook)
https://doi.org/10.1007/978-3-658-41230-2

This book is a translation of the original German edition „Sozialwissenschaftliche Datenanalyse" by Hartmann, Florian G., published by Springer Fachmedien Wiesbaden GmbH in 2022. The translation was done with the help of an artificial intelligence machine translation tool. A subsequent human revision was done primarily in terms of content, so that the book will read stylistically differently from a conventional translation. Springer Nature works continuously to further the development of tools for the production of books and on the related technologies to support the authors.

This Springer imprint is published by the registered company Springer Fachmedien Wiesbaden GmbH, part of Springer Nature.
The registered company address is: Abraham-Lincoln-Str. 46, 65189 Wiesbaden, Germany

Contents

Preface to the New Edition

It has been more than a decade since two of the now three authors – then still in Chemnitz – discussed the first considerations to write an introductory book on social science data analysis. Our aim was to create a short, but comprehensive book for our students, covering the basic logic of empirical social research and especially the individual practical steps of data analysis – starting with the question of data access, the presentation of the often underestimated efforts of data preparation and the description of the data and simple bivariate relationships, up to more complex and multivariate methods and their basic logic. It was important to us not to scare readers away with an overly formalized presentation, nor to lull them into a false sense of security by a detailed guidance in the different analysis programs. In 2012, the result of our considerations finally appeared and after a short time we were able to start a new edition in 2014, which included some additions and corrections, but retained the basic structure of the argument.

In this new edition, we have made some content adjustments. We refrain from discussing some basic methodical and methodological questions.[1] We also

---

[1] Some of these considerations apparently led to misunderstandings regarding the aim of the book and were understood as a rejection of not quantitatively oriented considerations and approaches. Social science research should proceed problem-oriented and use the most diverse approaches, without having ideological concerns. On the contrary, some of these seemingly rather off the empirical track considerations can even serve as a guideline for empirical research.

Supplementary Information: The online version contains supplementary material available at https://doi.org/10.1007/978-3-658-41230-2_1.

F. Hartmann et al., *Social Science Data Analysis*,
https://doi.org/10.1007/978-3-658-41230-2_1

deleted the chapter on factor analysis, as this topic is probably better suited for more advanced statistics textbooks than for an introduction. New is a detailed chapter on the introduction to the logic of significance tests. The task of this textbook is still to provide a hopefully well-readable and practice-oriented introduction to the necessary steps of data analysis.

This book is therefore not a general introduction to the methods of empirical social research—here, especially since the standard work by Rainer Schnell et al. (2018) there are also a multitude of different presentations in German language—and certainly not an introduction to statistics. The present book is not intended to be a new presentation of the supposedly same material or the well-known techniques and facts. Statistics in the social sciences should serve as a tool and is not an end in itself. Often, in corresponding books, the aim is to introduce the individual work steps of practical social research (Benninghaus 2006), to convey necessary statistical knowledge (Müller-Benedict 2011) or to explain the "fundamentals of social science statistics" (Diaz-Bone 2019, p. 11). Our long and not always easy experience in teaching and conveying corresponding analysis techniques raises doubts whether these goals, pursued in the many books, are really achieved.

A little provocatively and therefore exaggerating, one could say: This mediation of basic statistical knowledge—at least alone—is explicitly not the goal of the present book. It is also no coincidence or result of idiosyncratic preferences that the title does not read "Introduction to Statistics for Students of Social Sciences" or similar, but explicitly refers to the applications of data analysis in the social sciences in general and sociology in particular. The objective is not necessarily that the corresponding methods can be used independently, that is, that one is able to analyze data independently—that would be a nice side effect. More important seems to be that the critical understanding of empirical analyses is promoted, because there can ultimately hardly be any reasonable doubt that empirically based knowledge and thus a so-called evidence-based approach is gaining and should gain importance in almost all areas. Evidence-based knowledge, however, also implies that one is able to understand and possibly criticize the corresponding procedures. It is precisely this understanding that this book aims to foster.

To this end, we will deal in a first substantive chapter with the function of empirical social research and especially data analysis within a sociological research process. In accordance with the character of the book, this should not reflect a scientific-theoretical discussion, but rather demonstrate the usefulness of the methods to be explicated in the following chapters using a concrete example. Another rather preparatory chapter then deals with the question of where the data to be analyzed come from. Here, a deviating opinion from many textbooks is expressed. Because we ultimately strongly advise against motivating students to collect data independently. Already at this point in advance: Data collection is a time-consuming and costly undertaking that cannot simply be done en passant and must be prepared and carried out carefully and thoroughly. 'Some data' are not always better than 'no data'![2]

In Chap. 4 the first steps of data analysis are then presented, whereby data preparation is given a relatively large space, because in practice these steps not only prove to be usually very time-consuming, but also involve a multitude of decisions that are rarely made clear in their drama. Subsequently, data description and the search for simple, bivariate relationships will be the focus. Especially because empirical analyses are also becoming more and more widespread in public and political discourse and, thankfully, fact-based information and discussions are becoming increasingly important in many areas of life, descriptive and statistically not too elaborate methods are of great importance and should always be the first steps of data analysis.

The structure of the book, however, already reveals that one should not stop here, of course, the work. It is important in this context to distinguish between the foundation of the scientific analysis and statements and the presentation especially in the public discussion.

An outstanding example in this context is the study "The Shape of the River" by William G. Bowen and Derek Bok (1998), which probably refers to a metaphor by Mark Twain in the title. The two authors examine the long-term effects of the so-called affirmative-action policy for admission to higher education institutions in the United States. Here they work in the main part of the book almost exclusively with easy to understand bar charts and other graphics. These representations, however, are based on multivariate analysis methods, which are presented in detail in the appendix, which covers a good third of the book. This book, which is still very exciting in terms of content, is therefore an excellent example of how to practice 'public sociology' and at the same time work methodologically sound.

The second half of the present book then deals with exactly these multivariate methods, whereby two variants, linear and logistic regressions, are discussed in

---

[2] We therefore also refrain from addressing the many pitfalls of data entry and the need for data control resulting from it. However, this plea should not be understood to mean that corresponding steps should be ignored in the education of students. Only if one knows—at least theoretically—the problems of data collection and data entry, one can assess and possibly criticize the quality of corresponding studies—and this is usually an important task.

depth. Considerations on the problem of significance tests, on the logic of data analysis, as well as an overview of further and advanced analysis methods complement these presentations.

Two more important remarks, however, are still necessary in this preface: We still vehemently—and even with increasing emphasis—hold the view that the use of so-called syntax files for data analysis is simply without alternative, even if this word is sometimes negatively connoted today. We will explain this in more detail below. This has led in the previous editions to the fact that we had shown some examples in one or the other place. Now these attempts and efforts were unfortunately only sporadic and ultimately inconsistent. In view of the diverse data analysis programs with their individual advantages, but also difficulties, we have decided for this new edition to delete also these few examples, especially since such an introduction can not replace a manual of the individual data analysis programs. We therefore refrain completely from showing the corresponding syntax—the urgent admonition to proceed in this way remains.[3]

Finally, it was possible to win another author with Florian Hartmann, who can not only bring in new perspectives, for example on the logic of significance tests, but also ensure the continuity of this introduction.

References

Benninghaus, Klaus. 2006. *Einführung in die sozialwissenschaftliche Datenanalyse.* München: Oldenbourg. https://doi.org/10.1524/9783486837360.

Bowen, William G., und Derek Bok. 1998. The shape of the river. *Long-term consequences of considering race in college and university admissions. Princeton: Princeton University Press.* https://doi.org/10.1515/9781400882793.

Diaz-Bone, Rainer. 2019. *Statistik für Soziologen.* München: UVK.

Müller-Benedict, Volker. 2011. *Grundkurs Statistik in den Sozialwissenschaften. Eine leicht verständliche, anwendungsorientierte Einführung in das sozialwissenschaftlich notwendige statistische Wissen.* Wiesbaden: VS Verlag. https://doi.org/10.1007/978-3-531-93225-5.

Schnell, Rainer, Paul B. Hill, und Elke Esser. 2018. *Methoden der empirischen Sozialforschung.* München: Oldenbourg. https://doi.org/10.1007/978-3-658-20978-0_59.

---

[3]The analyses presented in this book were all performed with the program Stata. The syntax used for this can be viewed on the corresponding pages of the publisher Springer VS.

On the Task of Empirical Social Research and Data Analysis in the Sociological Research Process

2

One can of course doubt whether an answer to the question formulated in the title should and can be dealt with in a meaningful way on a dozen pages of an introductory textbook. Ultimately, one would have to answer this question in the field of epistemology and philosophy of science and the discussion about the right answer fills several shelves in corresponding libraries and countless pages in journals. Despite all the discussions and ramifications to be found there, a relatively simple understanding has prevailed in practice, which is called critical rationalism within the epistemological discussion (see Albert 1991 or the brief presentation by Schnell et al. 2018). Empirical research serves to carry out theoretically guided descriptions and tests of corresponding hypotheses.

Therefore, no attempt will be made at this point to present these or similar epistemological discussions even in outline, but rather a relatively pragmatic line of argument will be chosen, which is intended to show the advantages of such an approach in practice. To this end, a first step will take a look at two exemplary questions, before the further course will examine the probably increasing importance of this approach based on an analysis of the two leading German-language sociological journals, the Zeitschrift für Soziologie (ZfS) and the Kölner Zeitschrift für Soziologie und Sozialpsychologie (KZfSS).

The first of the examples presented here deals with the actual or supposed deproletarianization of football, at least in its media preparation. While football was long regarded as a worker and proletarian sport, football today seems to be accepted across all strata. If one looks for explanations for this change, one quickly points to the changed media presence and presentation, which led to a transformation into a show sport (see for the corresponding references Fürtjes and Hagenah 2011, p. 281 ff.) or even makes football part of the so-called event culture (see introductory Gebhardt et al. 2000).

F. Hartmann et al., *Social Science Data Analysis*,
https://doi.org/10.1007/978-3-658-41230-2_2

An important sociological argument in explanations is to point to changed conditions and structures. For this reason, Fürtjes and Hagenah (2011, p. 282) also formulate the so-called similarity thesis. It is assumed that "the deproletari-anization of football results essentially from the collective social rise of the West German society", i.e. from a general social change. A second argument aims at a change in the recruitment patterns of the football audience. Both assumptions sound plausible and one cannot really decide which consideration has a greater explanatory value. But it is precisely at such a point that empirical social research comes into play—or to stay in the picture of the example: onto the pitch.

In order to investigate such developmental hypotheses, it is of course neces-sary to have a data set that covers the corresponding period of investigation. For-tunately, in this example, one can resort to the so-called reader or media analysis (cf. Fürtjes and Hagenah 2011, p. 286 ff.), an annual survey on media use, which is based on a random sample and therefore allows a statistical evaluation (cf. more details in Chap. 7). As an indicator of interest in football, the regular use of the Kicker sports magazine is used—with all imaginable limitations. How has the composition of this clientele changed from 1950 to the 2000s—the underlying article was published in 2011? Figure 2.1 shows the proportion of readers from the working class and the salaried employees.

Looking at this figure, a deproletarianization is clearly noticeable in the 1960s. The proportion of the working class decreases, the proportion of the salaried

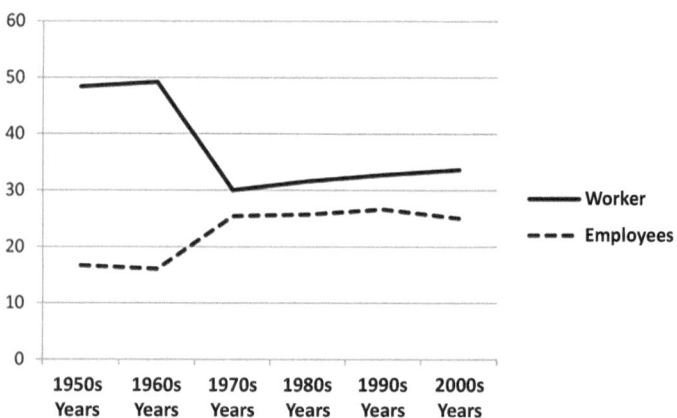

Fig. 2.1 Proportion of the working class and the salaried employees among the reader-ship of the Kicker (in percent). (Source: own representation based on Fürtjes and Hagenah 2011, p. 290)

employees increases, although hardly any changes can be found here since the 1970s. The change hypothesis formulated above is therefore empirically supported, even though this change occurs much earlier in the Kicker readership than is generally assumed based on the mediatization of football.

The research question is now whether this change can be explained by the change in the social-structural composition of the population—has the ratio of working class and salaried employees simply shifted accordingly—or whether additional factors—such as an opening of football through its media presentation—are necessary. To test this, a so-called logistic regression is calculated, which will be explained in more detail later in this book. Ultimately, it is about identifying influencing factors on the probability of reading the Kicker regularly. In Table 2.1 (cf. Fürtjes and Hagenah 2011, p. 294; for better clarity, some minor changes have been made) the so-called exposed β-effects can be found. These effects are relatively easy to interpret: effects greater than 1 indicate an increase in the chance of reading the Kicker, whereas effects smaller than 1 indicate a decrease in this chance. The distance from the value 1 indicates the size of these effects, whereby, for example, an effect of 1.17 means an increase in the chance by 17% and an effect of 0.83 means a decrease in the chance by 17%. Additionally, Table 2.1 also provides information on the significance level, the so-called p-values. These values are first of all to be understood as indicating how confident one can be that these are not random findings, but that there are really substantive results to report.

How are these results to be interpreted? Of particular interest are the interaction effects found at the end of the table. If there really had been a non-structurally caused deproletarianization of the Kicker readership over time, a negative effect—i.e. an effect below 1—for "year and workers" and positive effects—i.e. values greater than 1—for the other interaction effects with the occupational groups would have to be found. However, this is not the case! The chance for workers to belong to the core readership of the Kicker increases over time and so Fürtjes and Hagenah (2011, p. 296) also come to a clear result: "The analyses of the Kicker core readership produced a clear picture of the deproletarianization of football: The shift of the social focus of that readership results exclusively from the social structural change that has taken place in the past 50 years". The recruitment patterns within the changing social structure, on the other hand, are surprisingly stable or even rather speak for a slight proletarianization.

This example illustrates that an empirical observation, a puzzle, can be explained theoretically by different aspects and that empirical social research is needed to decide between the different explanations—as Fürtjes and Hagenah have shown very nicely.

Tab. 2.1 Social-structural determinants of regular Kicker reading. (Source: Fürtjes and Hagenah 2011, p. 294)

Factors	Effect
Occupation	
Non-employed	Reference value
Workers	0.90
Employees	1.27**
Civil servants	1.12
Self-employed	0.55***
Education	
Low education	1.90***
Medium education	1.72***
Higher education	Reference size
Income (in 1000 €)	1.13***
Age	
Under 24 years	3.56***
25–34 years	2.79***
35–44 years	2.64***
45–59 years	2.03***
60 years and older	Reference size
Gender	
Women	Reference size
Men	8.08***
Year	0.99***
Interaction effect	
Year and workers	1.01***
Year and employees	1.00*
Year and civil servants	1.00
Year and self-employed	1.01***

$^{*}p \leq 0.05; ^{**}p \leq 0.01; ^{***}p \leq 0.001$

The second example comes from one of the most respected scientific journals in the world, from the journal "Science" published by the American Association for the Advancement of Science. In February 2021, Bocar Ba, Dean Knox, Jonathan Mummolo and Roman Rivera published an article that deals with the role of diversity in the US police force (see Starr 2021 for a brief overview). Not only since the killing of George Floyd in May 2020, there has been a lot of discussion about how to prevent violence by police forces, with the dissolution of the previous police under the slogan "defund the police" being a quite widespread opinion. In the article briefly presented here, it is now examined on the basis of very detailed official records to what extent an opening of the police, which was exclusively "white" and male for a very long time, for members of other ethnicities or also the hiring of female police officers influences conflicts during patrols. To this end, it is important to consider that the composition of the respective patrols, but of course also their deployment times and areas, are important determinants for conspicuous behavior and thus also for police action. To control for these potential sources of error, these things were controlled in a very elaborate procedure: "To make valid comparisons, we assemble a panel dataset in which rows represent officer-shifts—roughly 8-hour patrol periods—and characterize officers' actions and their context. (…) In each of these 2.9 million patrol assignments, we measure officers' stops, arrests, and uses of force, whether they engaged in any of these activities or not. (…) This procedure greatly mitigates threats from self-selection" (Ba et al. 2021, p. 697). For these units, multivariate methods—so-called fixed-effect regressions—were then applied. The result shows that, for example, black police officers have significantly lower numbers of stops, arrests and especially uses of force than white police officers. Similar results are found for other ethnic groups and policewomen. "Taken together, these results strongly suggest that diversification can reshape police-civilian encounters" (Ba et al. 2021, p. 701).

Now, the different methodological questions and the possibilities of generalization are discussed in the article itself, but what is important in our context is that one can also examine and evaluate such highly political questions about ethnic conflicts and inappropriate police violence with the help of empirical methods—and this is the great merit of the study by Ba, Knox, Mummolo and Rivera.

It is not easy to stop here and not present further current or classic studies that demonstrate the appeal of sociology, the discovery of the new and the critical test of different theoretical considerations, i.e. simply the empirical-explanatory work. At least for a number of classic empirical works, one can refer to corresponding overview works (see, for example, Kaesler and Vogt 2007), otherwise one can probably hope that browsing through current journals will certainly reveal many interesting studies. For all those who do not want to face this sometimes tedious task themselves, at least with regard to a handful of classic studies, the work of Hunt (1991) can be pointed out, who impressively demonstrates the explanatory power of an empirically oriented social science using five examples.

The aim of this section is not to present the current discussion on any substantive problem area, but rather to make clear why one actually wants to do empirical social research at all—or sometimes also should or even must!

Sociology is an empirical science of experience[1]—sociological knowledge can usually not be gained without working empirically. With such a classification of the social sciences in general and sociology in particular, one is in good company: In the introduction of the first handbook of empirical social research, René König 1967 already makes clear that "sociology is only possible as empirical sociology", because sociology is the science of society and "science is ultimately only possible as empirical research" (König 1967, p. 3). Although René König also sees, of course, that other non-empirical methods exist and are significant, the central role of empirical observations cannot be expressed more clearly.

This primacy of empirical research is also increasingly evident in the corresponding academic journals. For this purpose, the KZfSS and the ZfS—as already mentioned above—the two still leading German-language academic journals were examined. Starting in 1970—or in the case of the ZfS in its first year of publication 1972—and then always in a five-year rhythm, all substantive works were first classified according to whether they work empirically. In a further step, it was recorded which basic approach is used for data collection, whereby here roughly between so-called qualitative and so-called quantitative methods were distinguished (see for a more detailed description and an analysis of the data up to 2010 the contribution by Kopp et al. 2012).

While empirical works were still in the minority at the beginning, they have been the increasingly growing majority since the 1990s. When one then looks at the basic approach, quantitative empirical studies gain the upper hand more and

---

[1] Even though definitional boundaries or even quarrels over definitions are ultimately unproductive, it should be noted here nevertheless that the classical division of the scientific world into humanities and natural sciences is permanently more than questionable. Tertium datur! That sociology is not a natural science, but that the ‚objects' of sociological analyses are meaningfully acting humans, whose actions and especially their intended and unintended consequences are the actual subject area of sociology—to use the classical Max Weber definition of sociology—makes our discipline precisely the exciting area that it really is. In most cases, however, it is the concrete actions of humans and not the reflection on intellectual products that are the object of analysis. Sociology is therefore not a humanities, but an empirically oriented social science. That the access to empirical reality and its interpretation is not a trivial task is the topic of this (and other) introductions to the methods of empirical social research and the corresponding data analysis methods.

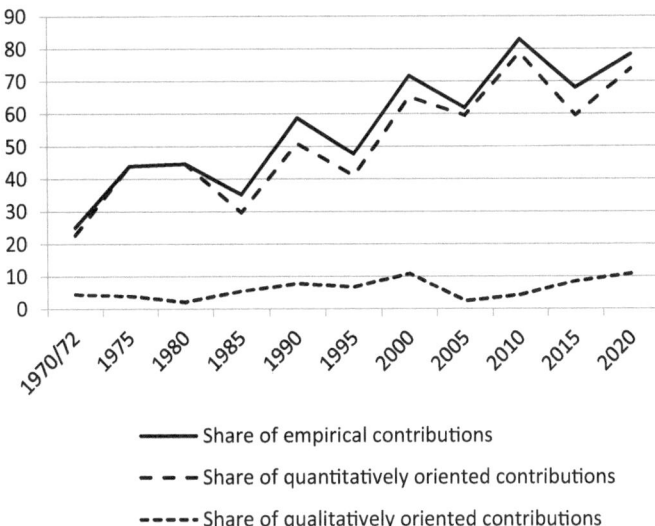

— Share of empirical contributions

– – – Share of quantitatively oriented contributions

----- Share of qualitatively oriented contributions

Fig. 2.2 Share of empirical studies in the German-language sociological literature (in percent). (Own representation)

more clearly. In Fig. 2.2 the corresponding shares of empirical studies of all published contributions for the two journals are shown.[2]

It is clearly visible that in the meantime about three quarters of all contributions in these two journals proceed empirically—whereby the quantitative methods that are also in the focus of this book clearly predominate. Less than 5% of the published contributions use qualitative methods, whereby this proportion has increased to around 10% in recent years. In a good three percent of the empirical works, both qualitative and quantitative methods are explicitly used together.

However, empirical research is not an end in itself and cannot be carried out without precondition. To conduct empiricism in the hope of being able to gain knowledge through mere—and be it ever so systematic—observation is a notion that is more than outdated in terms of the theory of science, even if the accusation

---

[2] This figure is based on the studies presented in Kopp et al. (2012), which were updated and supplemented for the years 2015 and 2020. Purely methodical and methodological contributions were not rated as empirical here. This would make the development even more clear.

of empiricism or in its intensified form of positivism is repeatedly formulated against a quantitatively oriented empirical social research. Therefore, it should be stated here clearly that both such an approach and the accusation just outlined are simply nonsensical. There is no theory-free empiricism (cf. Michotte 1963).

The starting point of many social science works is a concrete problem, a task that needs to be solved or a practical question that needs to be answered. In this respect, social and natural sciences are similar: "The natural sciences as well as the social sciences always start from problems" (Popper 1994, p. 17). Scientific work is the solving of puzzles and this is usually not a logical, but an empirical question.

Of course, there are completely different questions and (research) interests to be found. A first important step is the observation and description of empirical facts, developments and regularities. For example, Bienfait (2006) examines the practice of beatification and canonization of the various popes since 1592. For this purpose, a description of this development takes place first—but such a description never takes place purely without theory, but always along certain theoretical and substantive criteria.

Even pure description is thus oriented by theoretical and substantive categories and determined by theoretical explanatory mechanisms. Either one can explicitly face this task or these considerations flow unconsciously into the observation process—despite all subjective efforts to observe as unbiasedly as possible. This premise does not mean that the description of specific lifeworlds and a detailed sociography as they are historically found, for example, in the works of Friedrich Engels (Engels 1962) or more recently in the works of Roland Girtler on the lifeworlds of taxi drivers, prostitutes or smugglers (Girtler 2004,2006), are sociologically uninteresting. Although Heinz Maus (1967) classifies such studies in the prehistory of empirical social research, these works are of great interest and benefit for sociology students in the first semesters, as they show the diversity and colorfulness of the social world and thus possible social processes. Nevertheless, descriptions are not an unbiased representation of the world, but a picture of reality filtered by a multitude of theoretical premises.

After a first phase, it is necessary to look at the social world more systematically and subject the impressions possibly gained to a critical test. So-called factual experiments are needed. "It is always necessary to resort to the verification by means of experience" (König 1967, p. 8). Since social processes are usually neither deterministic nor monocausal, the question of the rules of these tests must be asked at the latest here.

As an intermediate conclusion, it can be stated that empiricism plays a decisive role in the process of knowledge acquisition.[3] It is precisely at this point that social science data analysis comes into play, which deals with the verification of theoretically formulated hypotheses about relationships.

The goal of sociology should be the explanation of real phenomena and not the production of a "flora of pseudoscientific phraseology" (Andreski 1974, p. 74), where only a jargon is cultivated. The sociological theory landscape, however, is diverse enough in the meantime to not only see the bad sides here. A certain caution is also necessary towards the attempts that strive for too much formalization without substantive support. The application is important, no overemphasis of methodology and thus no model platonism (Albert 1984).

Sociology is an empirically oriented science of experience. Sociological knowledge cannot—or more precisely: only in rare exceptional cases—be achieved by introspection or pure thinking. One almost always needs information about the world. Although there are discussions under the keyword 'constructivism' that the (social) world is a construction of the involved actors and thus does not exist in itself or objectively. Even if these discussions have a certain intellectual appeal, they are usually of little use for the practice of empirical social research and its various tasks. Ultimately, the primacy of empirical work applies in order to achieve real insights.

How one now arrives at these empirical facts, how one presents them understandably and how one examines the internal connection of the individual observed and measured facts in the most diverse empirical questions, with which strategy one carries out the corresponding data analysis—all these questions are to be dealt with in detail in the following sections. The aim is always to provide the readers with the necessary tools for a critical examination of empirical data, but also of already conducted empirical studies and their results. Only those who understand this tool at least and possibly can also use it themselves are able to critically examine existing work and thereby discover relationships.

---

[3] This is a fact that can already be found in the writings of Mao Zedong, which were very widely spread in the 1970s: "Have 'numbers' in your head. That means, one must pay attention to the quantitative side of a situation or a problem, must carry out a basic quantitative analysis (...). Many of our comrades still do not understand to this day to pay attention to the quantitative side of things (...) and consequently make inevitable mistakes" (Mao Tsetung 1972, p. 132 f.).

References

Albert, Hans. 1984. Modell-Platonismus. Der neoklassische Stil des ökonomischen Denkens in kritischer Beleuchtung. In *Logik der Sozialwissenschaft*, Ed. Ernst Topitsch, 352–380. Königstein: Athenäum.

Albert, Hans. 1991. *Traktat über kritische Vernunft*. Tübingen: Mohr.

Andreski, Stanislav. 1974. *Die Hexenmeister der Sozialwissenschaften. Mißbrauch, Mode und Manipulation einer Wissenschaft*. München: List.

Ba, Bocar A., Dean Knox, Jonathan Mummolo, und Rosman Rivera. 2021. The role of officer race and gender in police-civilian interaction in Chicago. *Science* 371:696–702. https://doi.org/10.1126/science.abd8694.

Bienfait, Agathe. 2006. Zeichen und Wunder. Über die Funktion der Selig- und Heiligsprechungen in der katholischen Kirche. *Kölner Zeitschrift für Soziologie und Sozialpsychologie* 58:1–22. https://doi.org/10.1007/s11575-006-0001-1.

Engels, Friedrich. 1962. *Die Lage der arbeitenden Klasse in England. Nach eigner Anschauung und authentischen Quellen*. Marx-Engels Werke (MEW), Bd. 2. Berlin: Dietz.

Fürtjes, Oliver, und Jörg. Hagenah. 2011. Der Fußball und seine Entproletarisierung. Zum sozialstrukturellen Wandel der Kickerleserschaft von 1954 bis 2005. *Kölner Zeitschrift für Soziologie und Sozialpsychologie* 63:279–300. https://doi.org/10.1007/s11577-011-0132-7.

Gebhardt, Winfried, Roland Hitzler, und Michaela Pfadenhauer, Eds. 2000. *Events. Soziologie des Außergewöhnlichen*. Opladen: Leske + Budrich. https://doi.org/10.1007/978-3-322-95155-7.

Girtler, Roland. 2004. *Der Strich. Soziologie eines Milieus*. Münster: LIT.

Girtler, Roland. 2006. *Abenteuer Grenze. Von Schmugglern und Schmugglerinnen, Ritualen und „heiligen" Räumen*. Münster: LIT.

Hunt, Morton. 1991. *Die Praxis der Sozialforschung. Reportagen aus dem Alltag einer Wissenschaft*. Frankfurt: Campus.

Kaesler, Dirk, und Ludgera Vogt, Eds. 2007. *Hauptwerke der Soziologie*. Stuttgart: Teubner.

König, René. 1967. Einleitung. In *Handbuch der empirischen Sozialforschung. Erster Band*, Ed. René König, 3–17. Stuttgart: Enke.

Kopp, Johannes, Juliane Schneider, und Franziska Timmler. 2012. Zur Entwicklung soziologischer Forschung. *Soziologie* 41:293–310.

Mao, Tsetung. 1972. *Worte des Vorsitzenden Mao Tsetung*. Peking: Verlag für fremdsprachige Literatur.

Maus, Heinz. 1967. Zur Vorgeschichte der empirischen Sozialforschung. In *Handbuch der empirischen Sozialforschung. Erster Band*, Ed. René König, 18–37. Stuttgart: Enke.

Michotte, Albert. 1963. *The perception of causality*. London: Metheun. https://doi.org/10.4324/9781315519050.

Popper, Karl R. 1994. *Alles Leben ist Problemlösen. Über Erkenntnis, Geschichte und Politik*. München: Piper.

Schnell, Rainer, Paul B. Hill, und Elke Esser. 2018. *Methoden der empirischen Sozialforschung*. München: Oldenbourg. https://doi.org/10.1007/978-3-658-20978-0_59.

Starr, Douglas. 2021. Study: Police diversity matters. Landmark analysis of 7000 police shows nonwhite and female officers make fewer stops. *Science* 371:661. https://doi.org/10.1126/science.371.6530.661.

On the Data Situation: Own Data Collection or Secondary Analysis

In many introductions to social science data analysis or to the methods of empirical social research, there are sections that deal with the creation of data sets and the input of independently collected original data. In the present book, this is explicitly omitted and this has the least to do with the always scarce space: The imparting of such basic knowledge is—with very few exceptions—usually simply superfluous. Normal users and especially students do not enter data—and that is a good thing![1]

As the various introductions to the methods of empirical social research and their individual steps (cf. for example Schnell et al. 2018; Schnell 2018) show, the number of pitfalls and problems that one faces when planning, conceptualizing and conducting an empirical—and by this we always mean: a quantitatively oriented empirical—study is so large that one should usually not even try to collect data independently within the framework of teaching research projects or even qualification works such as bachelor's, master's, but also doctoral theses. As a rule, the quality does not meet the professional standards (cf. the passionate plea by Schnell 2018).[2] The question posed in the chapter heading can therefore

---

[1] The figure "Little King" (cf. Munck 2015), which may still be known to many students from the evening Sandman, justified several actions with the statement "just to be on the safe side". In this tradition, we would like to paraphrase our remarks from the introduction for safety's sake, that this omission does not mean to leave out corresponding steps in the education of students. The knowledge of these problems of data collection makes the quality of empirical studies only assessable and, if necessary, criticizable.

[2] The only exception may be the creation of small sample data sets to understand the logic of certain statistical procedures and methods more precisely. However, such data sets can easily be created and manipulated by hand or with small syntax programs. Another

---

F. Hartmann et al., *Social Science Data Analysis*,
https://doi.org/10.1007/978-3-658-41230-2_3

usually be answered as easily as clearly: For most questions within the frame-work of qualification works, secondary analyses are sufficient despite all possible problems! It is not to be expected that one's own data collection will generate information that justifies these efforts (cf. once again Schnell 2018). It is an illu-sion that one has enough time within the framework of a teaching research project or even a bachelor's or master's thesis to carry out all steps of an empirical survey so carefully that one can expect reliable results. Data collection is not amateur work that can be done quickly!

A second and perhaps even more weighty argument is the fact that in the last decades various datasets have been made available for the interested academic public, whose potential has not been exhausted by far and which are increasingly available without much effort. Even if the publication policy of individual studies is certainly in need of improvement, there are rich opportunities in the various data archives to pursue scientific questions independently. Here, two important sources can be found: Almost all important institutions such as the Institute for Employment Research or the Statistical Offices of the Federal States or the National Educational Panel Study presented in more detail below have on the one hand so-called research data centers, where one can either analyze or obtain the information collected there. On the other hand, data sets have been collected for a long time, for example, in the GESIS Data Archiv for the Social Sciences and archived for reanalyses.

How important this is, is not only shown by the repeatedly flaring up discus-sion about the replication of research results (Freese and Peterson 2017), but also simply in empirical practice. Looking again at the development of the published contributions in the Kölner Zeitschrift für Soziologie und Sozialpsychologie and the Zeitschrift für Soziologie, there are clear developments. In Fig. 3.1 the pro-portion of empirical studies that are based on the analysis of existing, i.e. not self-generated or collected data sets, is shown for the period from 1970 to 2020.

---

exception could lie in the support of data that has to be collected anyway, such as the usu-ally legally required evaluations of courses, or in very clearly defined and thus also easy to verify questions, such as the content analysis of the Cologne journal and the Journal of Sociology presented above. However, anyone who looks at the multitude of examples in the corresponding textbooks (Schnell 2018; Porst 2014) may guess the potential for error in collecting data without sufficient qualification and preparation and ultimately also the financial resources required for conducting a reliable study. It cannot be explained in detail here that similar arguments ultimately also apply to studies that rely on qualitative meth-ods of empirical social research. Here, the efforts are different, but still efforts that usually exceed the possibilities of graduates.

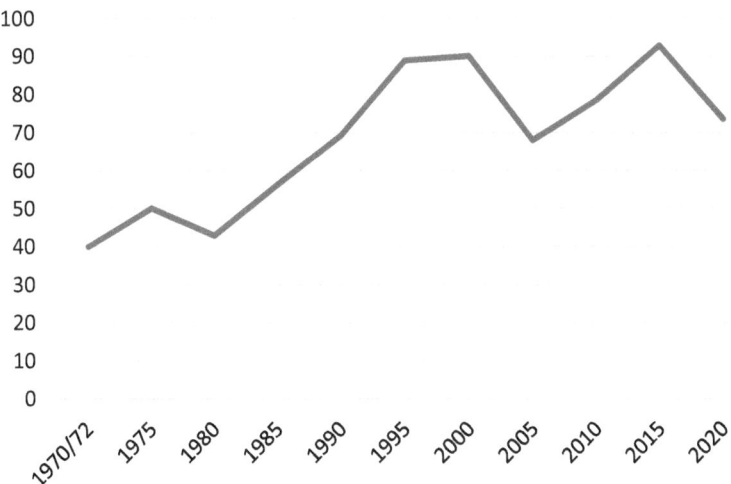

Fig. 3.1 Proportion of secondary-analytical empirical studies (in percent). (Source: own representation based on the data of Schneider and Timmler (2011) as well as their extension in 2021)

Here it is clearly visible that scientific publications no longer rely primarily on the analysis of primary data. Of course, there are also occasional drawbacks to using existing data, because certain things have not been or have not been collected as one would ideally imagine. Especially for qualification works, the disadvantages of an independent data collection clearly outweigh these limitations.

Here, no attempt will be made to give an overview of the diversity of data available in this way, such an undertaking would be doomed to failure from the outset. Rather, in the following, only some of the particularly interesting social science and especially sociological data sets will be sketched and briefly presented. All data have in common that they are provided free of charge or for low administrative fees for scientific analyses quickly.

German General Social Survey (GGSS, ALLBUS) The GGSS has been conducted since 1980 in a biennial rhythm as a replicative survey.[3] The GGSS is thus

---

[3]This section is based on the presentation of the GGSS/ALLBUS on the pages of the GESIS. There you can also find the modalities of how to obtain the data. We refrain from listing the easily searchable internet addresses here.

the oldest still regularly conducted social science data collection in the Federal Republic of Germany and according to its own statement a multi-thematic survey series on attitudes, behaviors and social structure of the population in the Federal Republic of Germany. The target population consisted until 1990 of the eligible population in the Federal Republic and since 1991 of the adult resident population in East and West Germany. The sample size was until 1990 around 3000 persons and thereafter about 3500 persons, whereby here persons in East Germany are overrepresented.

The sample is always based on a random selection, whereby the exact sampling method has changed several times over the years and can also be assessed quite differently (Sodeur 2007). Since the year 2000, the selection is based on a population register sample and takes place as a computer-assisted personal interview (CAPI). The survey planned for the year 2020 was postponed due to the pandemic.

In the individual GGSS data, different thematic focuses are in the center. This ranges from the topic of social inequality, ethnocentrism or social capital in the year 2010 over leisure activities and media use in the year 2014 to national pride, right-wing extremism and social networks in the GGSS 2018. In addition to these content-related focuses, demographic information is collected in each wave (cf. Hoffmeyer-Zlotnik 2015 as well as the information on the homepage of GESIS), which include general biographical data, but also information on income and current living conditions as well as marital status.

Since 1986, the GGSS has been supplemented by the questionnaire program of the so-called International Social Survey Programme (ISSP). The ISSP also pursues a specific survey focus, starting with social networks and support potentials over family and changing gender roles to leisure, sport and religion in the year 2018. While at the beginning these data were collected only in four countries, today people in almost fifty countries are surveyed on these topics and thus offer the opportunity for extensive international comparisons.

In the GGSS, there is also the possibility to supplement the individual data with characteristics of the social near environment, more precisely with sensitive regional data and small-scale geodata. This research perspective is promising, because the living conditions in the Federal Republic of Germany vary not only along the traditional socioeconomic dimension of social inequality, such as education. In addition, depending on the research question, it makes a significant difference whether people live in rural or urban regions or from which area of Germany they come (Bayer 2010). However, regional data are not available in the normal GGSS data sets for data protection reasons. However, if a special agreement is made with the Secure Data Center, the user gets access to so-called

"restricted use files" (cf. again the corresponding pages at GESIS). For the technical implementation of such analyses, first hints can be found in Chap. 9.

For scientific purposes, the GGSS data can be obtained online after a short
registration. For a small fee, a corresponding CD-ROM can also be purchased.
The number of content analyses with this data set is large—on the internet pages
of the GGSS at GESIS you can find a very extensive bibliography.

**Studies in the field of family sociology: Family Survey, Generation and Gen-
der Survey and pairfam** Especially in the field of family research, there have
been large data sets for a long time. A first study in the field of family development is the so-called Family Survey, whose first wave was collected in 1988 and
in which more than 10,000 people between 18 and 55 years of age were interviewed. In 1991, an additional 2000 people in East Germany were surveyed. The
focus of the study was the diversity and change of family forms, the network
structures of family and kinship, the dynamics of partner relationships, issues in
the field of fertility, i.e. births and upbringing of children, and finally the occupational careers with their effects on family life. Around 5000 people in West Germany were interviewed again in a second wave in 1994. The follow-up survey in
East Germany was not feasible for technical reasons. Here, but also in West Germany, the panel sample was supplemented by around 5000 new interviews. The
third wave of the Family Survey from the year 2000 comprises a total of around
10,000 interviews again, with only around 2000 respondents being available as a
panel sample.

Despite the just indicated problem of the samples, which certainly only allow
relatively limited descriptive statements, the Family Survey is a good data source
for exploring familial processes, as a variety of interesting constructs—such as
the integration into social near environments and much more—were collected.
The data of the Family Survey and some accompanying studies are available from
GESIS.

The "Generations and Gender Survey" (GGS) is institutionally embedded in
the "Generations and Gender Programme" (GGP) of the "United Nations Economic Commission for Europe" (UNECE). Within this project consortium,
several survey waves (with panel design) on the topics of fertility, partnership
development and intergenerational relations were and are conducted in various countries. These currently include, besides Germany, for example Australia,
Belgium, Bulgaria, Estonia, France, Georgia, Italy, Japan, Lithuania, the Nether-
lands, Norway, Romania, Russia, the Czech Republic and Hungary.

The German part of the Generations and Gender Survey is a representative
sample of—in the first survey wave—more than 10,000 German-speaking people

aged between 18 and 79 years who live in private households in Germany. The survey is conducted on behalf of the Federal Institute of Population Research (BIB) and is intended as a successor to the Family and Fertility Survey (FFS) from 1992. The first wave of the GGS took place in 2005, and in 2006 a second survey was conducted among the Turkish population living in Germany. In the late 2000s, the respective second waves were collected, and the third wave was planned for 2020, but had to be postponed in part due to the Corona pandemic.

The aim of the GGS is to obtain current data on family relations in industrialized countries through a multidisciplinary, retrospective, prospective and internationally comparative study. The survey is based on theories from different social science disciplines. Important components of the survey program are the respondent's own family situation at the time of the interview, family-related events in the past, intentions of the respondents regarding important demographic behaviors such as partnership formation, fertility behavior or leaving the parental home, as well as the socio-economic context of the respondents such as employment and education, income and wealth, health, social networks, values and attitudes. An important innovation of the GGS is also the detailed recording of generational and gender relations, which are supposed to contribute to the explanation of individual demographic behavior. To this end, values and attitudes of the respondents regarding the relationship between the sexes, questions about the relationship between the generations (for example, frequency of contact, money transfer, emotional support), as well as questions about the division of labor in the household and the decision-making and use of household income between the partners are collected.

The scientific strengths of the GGP or GGS project consist overall in the longitudinal perspective and especially in the international comparative perspective by collecting largely comparable indicators in a number of countries. Further information on this data set can be found on the homepage of the Federal Institute of Population Research and the Generation and Gender Programme.

Since 2008, the German Family Panel (panel analysis of intimate relationships and family dynamics, pairfam) has been conducted in the Federal Republic of Germany. The German Family Panel is a multidisciplinary longitudinal study that takes into account sociological, psychological and pedagogical aspects of the research of partnership and family forms.[4] The study

---

[4]The sketch of these data is also based on the presentations on the project's websites, which can be easily found under "www.pairfam.de". Here you can also find the contact details for ordering the data. There you can also find information about the development of the instruments used and a number of working papers and documentation.

started in 2008 with the survey of about 4000 people each from the birth cohorts 1971–1973, 1981–1983 and 1991–1993. In the eleventh wave, the birth cohort 2001–2003 was added as a refreshment. The panel was originally designed for a total of 14 years, resulting in a cohort-sequence design. At the moment, the data of the first twelve waves as well as a special survey on the Corona epidemic are available. The study follows a so-called multi-actor perspective, which means that in addition to the target or anchor persons, their partners as well as their parents and possibly existing children are also interviewed at regular intervals (see also Huinink et al. 2011 besides the website of the project).

The pairfam sample is based on a two-stage random procedure, in which first municipalities were randomly selected and then an address sample was generated in the municipalities with the help of the residents' registration offices. The persons selected in this way were then finally interviewed in so-called face-to-face interviews. A total of five thematic focuses are collected:

- the development of partnerships and especially the processes of getting to know the partners, establishing and shaping the couple relationship as well as possibly also the separation; the expectations of partnerships as well as other aspects of the quality and stability of partnerships
- the decision-making processes for family formation and expansion and especially the timing, spacing and stopping of births in the life course, the desire for children and the number of children as well as closely related to this the sexuality and contraception
- the relationship quality and transmission processes between generations, intergenerational transfer services of material and immaterial nature, family norms and expectations of the parents
- the educational goals and competence of the parents, educational behavior and childcare, child development processes, parent-child relationships in the family system
- the modeling of contextual influencing factors on processes of partnership and family development through network integration and the consideration of external regional indicators.

The Socio-Economic Panel (SOEP) The SOEP is a representative longitudinal survey of private households in Germany, which is conducted annually since 1984 with the same individuals and families.[5] Since 1990, the study has also been

---

[5] Since some people drop out of the panel survey over time (so-called panel mortality), the SOEP contains only partially the same individuals over longer periods of time. To prevent

extended to the territory of the former GDR. In total, about 15,000 households and 30,000 respondents participate in the SOEP. The anonymized data are prepared, documented and made available to the scientific community in Germany and abroad by the SOEP group at the German Institute for Economic Research (DIW) for a small user fee for purposes of research and teaching.

The Socio-Economic Panel is mainly used in the social and economic sciences, but also in the context of social reporting and policy advice. The data set contains information on the following main topics, which are collected continuously:

- Demography and population
- Work and employment
- Income, taxes and social security
- Family and social networks
- Health and care
- Housing, equipment and services of private households
- Education and qualification
- Attitudes, values and personality
- Time use and environmental behavior
- Integration, migration and transnationalization
- Survey methodology

The survey program of the SOEP has been and is continuously adapted to current developments. By default, individuals in private households aged 17 and over are interviewed. Over time, various extensions of the survey population and the questionnaire have been added:

- Since the survey year 2000, a youth questionnaire has been collected from the 16–17-year-old household members.
- From 2003 onwards, mothers of newborns answer their own questionnaire. From 2005 onwards, the parents of 2–3-year-old children (and starting in the years 2008, 2010 and 2012 the children of 5–6-year-old, 7–8-year-old and 9–10-year-old children) are also surveyed. Thus, the SOEP also represents a cohort study since the birth cohort 2003.

---

a too strong shrinking of the initial sample, refreshment samples are drawn regularly (so far in the years 1998, 2006, 2011, 2012 and 2017).

- As part of the study "Families in Germany" (FiD), family policy relevant target groups such as single parents, multi-child families and families with low household income were integrated into the SOEP in 2014.
- Since 2012, scientists also have the opportunity to contribute their own proposals for their respective research project to the survey program within the framework of the "SOEP Innovation Sample".
- The effects of the corona epidemic are examined with a special survey (SOEP-CoV).

The data of the SOEP are of special interest in two respects: On the one hand, they make possible annual situational analyses and thus contribute to the measurement of social change and on the other hand, they provide an important basis for testing theoretical explanatory approaches for the behavior of individuals or groups (cf. Schupp 2009). The scientific strengths of the SOEP can be summarized in the following points:

- Longitudinal design or panel character: Within the framework of quasi-experimental designs, for example, it can be examined how people change in the course of important life events—unemployment, birth of a child, moving from East to West Germany, etc.
- Household context: All adult household members are interviewed and information about the children living in the household is also collected; this provides, among other things, information about both spouses, for example their similarity in terms of educational level, values or leisure behavior.
- Possibility of regional comparisons and use of small-scale context indicators: Under certain data protection conditions, the DIW provides regional data. The SOEP can also be linked with other administrative data or survey data (see Goebel et al. 2019).
- Over-proportional immigrant sample: The SOEP is currently the largest repeated survey of foreign persons in the Federal Republic, with the sample covering household heads with Turkish, Spanish, Italian, Greek and former Yugoslav nationality. Currently, the SOEP comprises four immigrant samples (1984, 1994, 2013, 2015) as well as a special refugee survey in 2016.

Since the SOEP has been running for over 35 years, a wide range of birth cohorts are represented. This opens up a variety of analytical possibilities also in medium- and long-term temporal perspective: How do early life events affect later life course? How do intergenerational mobility and transmission processes work? What are the short- and long-term consequences of institutional change or political changes (with reunification in 1990 as probably the most important example)?

Overall, the SOEP represents one of the most important and widely used data bases for sociological research. Thus, in the specially maintained database "SOEPlit" by the DIW, there are now over 11,000 entries of publications that are based on a secondary analysis of the SOEP. In many cases, these are contributions in leading peer-reviewed journals. Further information can be found in the extensive web offer on the pages of the DIW or in Schupp (2009) or Goebel et al. (2019).

National Educational Panel Study (NEPS) The NEPS combines different disciplines and pursues the goal of collecting longitudinal data on educational processes, educational decisions, educational returns and competence development in different contexts and across the life course. The resulting data is provided by the LIfBi to the scientific community in the form of Scientific Use Files. Three different data access modes are distinguished, which are associated with different degrees of anonymization and corresponding information (e.g. regional information) as well as security precautions for data use.

In terms of content, the NEPS offers the possibility to analyse educational processes from the perspective of sociology, educational science, psychology, demographics or economics. The framework concept of the NEPS includes eight educational stages, each of which focuses on critical transitions in the educational process, as well as six content dimensions along which educational trajectories can be described and analyzed. The following stages are distinguished:

- Stage 1: Newborns and early childhood education
- Stage 2: Kindergarten and transition to elementary school
- Stage 3: Elementary school and transition to a type of secondary school
- Stage 4: Paths through secondary school and transitions to upper secondary school
- Stage 5: Upper secondary school and transitions to higher education, vocational training or labor market
- Stage 6: Transitions to vocational training and to the labor market
- Stage 7: Higher education and transition to the labor market
- Stage 8: Adult education and lifelong learning

The following six content dimensions ("pillars") are covered for each stage:

- Competence development in the life course: This focuses on the longitudinal assessment of domain-specific and cross-domain competencies such as language competence, mathematical and scientific competence, and computer- and internet-based competence.

- Educational processes in life course-specific learning environments: Learning environments offer the opportunity to acquire certain skills and competencies. The learning environments in NEPS include formal contexts such as school, training place or university, non-formal environments such as child and youth welfare, clubs or religious communities, and informal environments such as family and peers or media.
- Social inequality and educational decisions in the life course: This dimension involves the analysis of class- and gender-specific educational decisions. The central question is why and to what extent group differences persist even when similar achievements are present and what role third variables such as success expectation or the costs anticipated with education play in this.
- Educational attainment of persons with a migration background in the life course: Beyond social inequality, the migration background is associated with further specificities or contexts that can explain an independent share of the variance of educational and labor market success. In particular, the acquisition of first and second language is taken into account here.
- Educational returns in the life course: This dimension examines educational outcomes and their causes. The concept of educational returns includes, for example, political and social participation as well as physical and mental health.
- Motivational variables and personality aspects in the life course: Within this pillar, the question is pursued to what extent social and personality psychological variables affect the educational trajectory and to what extent the latter in turn has a feedback effect on such variables. Important characteristics in this context are, for example, learning motivation, self-assessed self-esteem, and general interest orientations.

The data collection is carried out according to a multi-cohort-sequence design with six NEPS starting cohorts (SC1–SC6), which comprise a total of more than 60,000 target persons:

- SC1: Newborns, i.e. infants in their first year of life
- SC2: Four-year-old kindergarten children
- SC3: Pupils in grade 5
- SC4: Pupils in grade 9
- SC5: First-year students
- SC6: Adults (birth cohorts 1944–1986).

For the year 2022, the start of a new cohort is planned: In SC8, students of grade 5 will again be surveyed and tested. Furthermore, data from additional surveys will be provided, such as those on the Corona pandemic (NEPS-C). In addition, a small number of items from the scientific community will also be included in the survey. Under certain conditions, strongly anonymized versions of the data sets can also be used for teaching purposes as part of a currently (in 2021) ongoing pilot phase. An international NEPS conference takes place annually, where the own research results can be presented and discussed. Further information on the NEPS study and data access can be found on the NEPS websites. In particular, the user-friendly tool NEPSplorer is recommended. It allows a quick and uncomplicated search for collected variables, taking into account the cohort and survey wave.

References

Arbeitsgruppe Regionale Standards. Ed. 2013. *Regionale standards*. Köln: GESIS. https://doi.org/10.21241/ssoar.34820.

Bayer, Hiltrud. 2010. Regional tief gegliederte Daten im Bereich Bildung, Familie, Kinder und Jugendliche. RdJB Recht der Jugend und des Bildungswesens. *Zeitschrift für Schule, Berufsbildung und Jugenderziehung* 2:176–195. https://doi.org/10.5771/0034-1312-2010-2-176.

Freese, Jeremy, and David Peterson. 2017. Replication in social sciences. *Annual Review of Sociology* 43:147–165. https://doi.org/10.1146/annurev-soc-060116-053450.

Goebel, Jan, Markus M. Grabka, Stafan Liebig, Martin Kroh, David Richter, Carsten Schröder, and Jürgen Schupp. 2019. The German Socio-Economic Panel Study (SOEP). *Jahrbücher für Nationalökonomie und Statistik* 239:345–360. https://doi.org/10.1515/jbnst-2018-0022.

Hoffmeyer-Zlotnik, Jürgen Hans Peter. 2015. *Standardisierung und Harmonisierung soziodemographischer Variablen*. Mannheim: GESIS – Leibniz-Institut für Sozialwissenschaften (GESIS Survey Guidelines). https://doi.org/10.15465/gesis-sg_012.

Huinink, Johannes, Josef Brüderl, Bernhard Nauck, Sabine Walper, Laura Castiglioni, and Michael Feldhaus. 2011. Panel analysis of intimate relationships and family dynamics (pairfam): Conceptual framework and design. *Zeitschrift für Familienforschung* 23:77–100. https://doi.org/10.20377/jfr-235.

Munck, Hedwig, 2015: *Der kleine König*. Hamburg: Ellermann.

Porst, Rolf. 2014. *Fragebogen. Ein Arbeitsbuch*. Wiesbaden: Springer VS. https://doi.org/10.1007/978-3-658-02118-4.

Schneider, Juliana, and Franziska Timmler. 2011. *Zur Entwicklung soziologischer Forschung. Eine quantitative Inhaltsanalyse der „Zeitschrift für Soziologie" und der „Kölner Zeitschrift für Soziologie und Sozialpsychologie"*. Bachelor-Arbeit an der Technischen Universität Chemnitz.

Schnell, Rainer. 2018. *Methoden standardisierter Befragungen.* Wiesbaden: Springer VS. https://doi.org/10.1007/978-3-531-19901-6.

Schnell, Rainer, Paul B. Hill, and Elke Esser. 2018. *Methoden der empirischen Sozial-forschung.* München: Oldenbourg.

Schupp, Jürgen. 2009. 25 Jahre Sozio-oekonomisches Panel – Ein Infrastrukturprojekt der empirischen Sozial- und Wirtschaftsforschung in Deutschland. *Zeitschrift für Soziolo-gie*38:350–357. https://doi.org/10.1515/zfsoz-2009-0501.

Sodeur, Wolfgang. 2007. Entscheidungsspielräume von Interviewern bei der Wahrscheinli-chkeitsauswahl: Ein Vergleich von ALLBUS-Erhebungen. *Methoden, Daten, Analysen* 1:107–130.

The First Steps of Data Analysis: Preparation, Data Description and Bivariate Relationships

As has become clear, the main focus of social science data analysis is on the examination of the relationships between interesting facts or variables. In doing so, it usually requires little sociological imagination to suspect relatively complex and diverse causal relationships. The use of multivariate analysis methods, which are also the focus of this introduction, represents an attempt to take this fact into account and to shed some light on the thicket of theoretically possible and meaningful, but not always simultaneously true, relationships. Despite this fact, multivariate methods should not be at the beginning of an empirical analysis. It is indispensable in a first step of data analysis to take a closer look at the available information, the data set, to get to know the individual variables and their distributions more precisely. Here, two points have to be distinguished: data preparation and data description. One has to first bring the data into a meaningful and interpretable form and, for example, create summary indices and then examine these variables, preferably together with the underlying items, more closely. This descriptive analysis is a first important step, which also often allows to detect errors in data preparation. Only then is it possible to examine relationships between variables or better constructs.

This also describes the structure of this chapter: In a first step, some problems of data preparation will be discussed, before the description of the data with the help of various graphical methods and measures can be presented. Finally, depending on the corresponding scale level of the captured constructs, it will be examined whether there is a relationship between two variables. For all those who want to analyze social science data themselves, it should be pointed out that they have to bring a not too low level of frustration tolerance, because the work steps

F. Hartmann et al., *Social Science Data Analysis*, https://doi.org/10.1007/978-3-658-41230-2_4

at the beginning of this chapter usually prove to be much more time-consuming and error-prone than the actual interesting multivariate data analysis. The treatment of these individual steps in introductory books, if such remarks can be found at all in the literature and these enormously important things are not simply sovereignly skipped, is empirically not related to the work involved. Here, too, one can almost only point out how important these steps are, without being able to go into detail on the various problems and pitfalls. These problems with data preparation arise not only with primary surveys, but also with secondary analyses, as will become clear in the following using the data of the surveys of the GGSS/ALLBUS, which are the focus here.[1]

4.1   "A Long and Winding Road"—On the Difficulties of Data Preparation[2]

Almost regardless of which data are analyzed, whether own data were collected and coded or existing data were used for reanalysis, the following applies: The facts of interest in terms of content and theory are usually not available in such a way that they can be analyzed easily and simply. As a rule, the data have to be prepared, new variables have to be constructed or even the data set has to be changed in its basic structure. Based on the questions about gender role orientation or family images, which are initially the focus of this text, some examples will be given of how certain variables can be formed and what difficulties can arise.

---

[1] We use here mainly the data of the ALLBUS/GGSS from the years 2016 and partly the survey from the year 2018. The version of the cumulative GGSS archived at GESIS allows, however, in some cases to look at longer periods of time. The cumulative GGSS, which contains all available data harmonized, can be found in the GESIS data catalog under the number 5276. Not only because of the oversampling in the new federal states, the data of the GGSSS would ultimately have to be weighted. However, we largely refrain from doing so in order to make the calculations more comprehensible. However, great caution is required for substantive interpretations.

[2] Those who compare the effort and problems of data preparation with those of data analysis will be able to understand the quote hinted at in this headline. Depending on the elegance of the programming, the preparation of existing data can make up a good 95% of the corresponding syntax files. Overall, however, the sentences of Bertolt Brecht (1977, p. 960) still apply: "When I returned/My hair was not gray yet/Then I was glad./The troubles of the mountains lie behind us/Before us lie the troubles of the plains".

The starting point of the following explanations are three questions from the GGSS questionnaire, which capture the gender role orientation and reflect the attitude towards a rather traditional or conservative family image.[3]

- The first item is worded as follows: "For a woman, it is more important to help her husband with his career than to pursue a career herself". We will briefly call this item "career" and refer to it in this way.
- The second item is formulated as follows: "It is much better for everyone involved if the man is fully involved in his professional life and the woman stays at home and takes care of the household and the children". We will call this item "division of labor".
- And the third item considered here is: "A married woman should give up working if there are only a limited number of jobs available and if her husband is able to provide for the family". Here we will use the term "housewife" as a keyword.

For all three items, a response scale is available that has four levels, starting with "strongly agree", "somewhat agree" and "somewhat disagree" to "strongly disagree". In addition, there is also the possibility to answer with "don't know" or "no answer". The corresponding answer options are coded with the numbers 1 to 4 or with the negative numbers for the non-content reactions.

If one wants to form a scale of gender role orientation from these individual items, which represents a traditional or conservative family image, two steps have to be taken: In a first step, the corresponding information has to be recoded so that higher values also correspond to a more conservative family image and lower values indicate a rejection of this idea. In addition, only valid answers have to be considered, so that persons who have not expressed an opinion here do not enter the analyses with the numerical values for these so-called missing values.[4]

---

[3] The exact wording of the question and answer options are documented on the GESIS homepage. The corresponding questions can be found in the GGSS surveys of the years 1982, 1991 and from 1992 every four years and thus also in the year 2016. On the GESIS pages you can also find the corresponding method reports of the individual surveys as well as a number of other important information.

[4] If one neglects this ultimately trivial fact, fundamental errors can arise. In an analysis of partner fidelity (Munsch 2015; cf. especially Munsch 2018), the persons who did not provide information on the number of sexual partners were accidentally included in the analyses with a very high value, since the corresponding analysis program Stata stores missing values as high numerical values, which have to be excluded separately in the analysis. In the article mentioned, however, it was assumed for married persons that if the values of

In the case at hand, we have therefore reversed the values, high numerical values now indicate a strong agreement with the individual questions. In addition, the answers "don't know" or "no answer" would be marked as missing values, so that they are not considered in the substantive analyses.

In a next step, one can look at the distributions of the three items separately. This task is simple on the one hand, but on the other hand one can make a number of errors or at least take such inconsiderate steps with these frequency distributions and simple graphs that the actual goal—to reflect an understanding of empirical facts—is not achieved. We will go into this in detail in the further course of the chapter.

Before that, however, we would like to stay a little longer with the problem of data preparation, because ultimately we are only interested in the rarest cases in the concrete answers to individual questions, but we suspect certain causal relationships in the real world and assume, for example, that ideas about family life have an influence on behavior on the one hand and on the other hand—and this is ultimately a sociological axiom—are influenced by certain social situations, the class position, socialization, social status and many other things. These ideas can only be captured in the rarest cases with a single item, but must be mapped by different questions, which then reflect together as well as possible these basic factors.[5] This is also the case in the here interesting case of gender role orientation or conservative family image. As sensible as it may be to understand the three items mentioned above as equivalent measurements of a conservative or traditional

---

the variable "number of sexual partners in the last 12 months" were higher than 1, marital fidelity was not very strictly observed. This error would have been obvious in a simple descriptive examination, but in the present case a dichotomous variable—fidelity yes or no—was simply formed. Here, too, the meaningfulness of simple and descriptive methods becomes apparent.

[5] In individual, but rare cases, this argument may be wrong. For example, voting intention can be captured quite well with one item, the so-called Sunday question, regardless of the fact that election research has to cope with many other problems such as social desirability. A few years ago, Sara Konrath and two colleagues (Konrath et al. 2014) reported that the construct of narcissism, i.e. self-love or self-admiration, does not have to be measured as usual in psychological research by longer item batteries, but that a simple question can be asked, which reads as follows: "To what extent do you agree with this statement: I am a narcissist (Note: The word narcissist means egotistical, self-focused, and vain)" (Konrath et al. 2014, p. 3). Corresponding tests show that this simple measurement has the same quality as more complex measurements of the construct of narcissism. However, it is likely that such simple measurements are the exception.

family image, one should not simply determine this—a procedure that goes by the beautiful name of per-fiat measurement—or dogmatically fix it. One can, should and ultimately must test the thesis formulated here. However, since this test requires statistical methods that we just want to introduce or that go beyond the approach of this book, we ask for a little patience and trust in our assurance that the three items capture a uniform content dimension.[6]

The question now is, however, how these three individual items are combined into a uniform construct? Here different approaches are possible, which should be briefly outlined. On the one hand, it is possible to use the statistical methods just used, but only hinted at and not explained, the principal component analysis, and to determine corresponding factor values. This can be easily requested from all known analysis programs. The corresponding values are usually very finely divided, but are only calculated for the cases that have also provided meaningful answers for all items. Usually we lose some cases this way, because missing values due to nonresponse are common in empirical social research. In addition, the corresponding new variables are standardized, which means they have a mean of 0 and a standard deviation—what that is, we explain in a moment—of 1. Not everyone sees an increase in comprehensibility in such a standardization.

In order to optimize the number of analyzable cases and to keep the metric at the known four-level scale, we will therefore take a different approach in the following and not consider the factor values, but form a mean value index. However, decisions are also necessary here: What should happen if a person has no opinion on one of the three items or does not want to say it and thus only two of the three items have been answered? And what should happen if only one meaningful answer was given? Well: There is no universally valid answer to this, but we would tend to use as much information as possible and therefore also keep cases in the analysis that only answer one item.[7] In addition, meaningful variable

---

[6] For the impatient or mistrustful readers—and it should be explicitly pointed out that doubt is a very positive quality in the sciences—it should be mentioned here that in a simple exploratory factor or principal component analysis, which ultimately examines the correlations of the individual indicators and wants to extract common factors, these three items really lie on a single factor, which explains almost 63% of the total variance. The factor loadings of the three items range from 0.74 to 0.83, a corresponding analysis yields a Cronbach's α value of 0.70 and thus a sufficiently good reliability coefficient. The patient readers who nevertheless followed these remarks should be assured summarily that the assumption that these three items capture a common construct is probably correct.

[7] When presenting multivariate methods, we will see that such peculiarities can be taken into account in the analyses, for example by forming appropriate control variables and including them in the analysis.

and value names should be assigned that really reflect the intended content of the items.[8]

One could now easily get lost in the depths of data preparation and variable construction at this point and discuss the various problems and their solutions. As sensible as such a procedure may be, it would ultimately be of little benefit, because depending on the structure of the available data set and especially the problem, new questions arise that cannot all be comprehensively addressed. Nevertheless, at least one problem should be pointed out here, which can occur with different substantive questions and should therefore be briefly discussed here. This concerns the restructuring of the data set.

Often, the units that constitute the cases of the data set do not represent the analytically interesting units of analysis. In most cases, the data sets are structured in a case-oriented way. Each row of the data set corresponds to an interview, i.e. the information of a specific person. However, the focus of the research interest can be completely different things. This will be discussed using two examples:

- In many studies, the so-called partnership and marriage biography is collected, i.e. questions are asked about the beginning and possible end of romantic relationships and marriages (and many other things). If one is interested in the duration of relationships and possible determinants of stability, a person can certainly contribute experiences from different biographical phases. However, the data of one person are arranged within one row. To perform the outlined analyses, a new data set must first be formed that presents the respective information uniformly and, for example, captures the beginning of the relationship in a variable in addition to a counting variable that reflects the ordinal number of the relationship. Technically, a data set in wide format is converted into a data set in long format.[9]

---

[8] As obvious as this may sound, it is often violated in practice and a variable "spatial proximity to a person" measures, for example, the distance.

[9] An equivalent problem exists, for example, in the analysis of intergenerational relationships. If one conducts such studies from the perspective of the children, several parents can obviously be named—in addition, there are the relatively common step- or possibly patchwork families. Here, too, one and the same child can form different cases of the data set, since it can have relationships to fathers and mothers, to biological and social parents. The focus of interest is on the individual relationships, in the data set then one relationship forms one row. However, it must be taken into account that this creates so-called clumping effects that affect the significance tests and therefore have to be taken into account.

- Especially with the topic of marriage, it is a perhaps trivial fact that the course of relationships depends on the characteristics and behaviors of both partners. For this reason, there have been increasing efforts in recent times to either interview both partners or at least take into account the characteristics of both partners in an analysis (see, for example, Hahn et al. 2019). In most cases, however, even with such surveys, the interviews of the partners are separate and one is forced to merge the information of both partners, so that a data set is created that comprises a couple as the unit of analysis in one row.

In summary, it is often necessary to reorganize the data structure for an appropriate answer to the research question of interest. Here, too, the utmost care is necessary, because especially with complex preparations, the risk is high of performing incomplete or erroneous transformations. The use of syntax files, which was already justified at the beginning, is without any alternative, if one wants to keep the work comprehensible for oneself, but also for possible replications and extensions. Too much elegance of programming rarely goes along with a high intelligibility and is therefore certainly not the highest goal. Ultimately, syntax files should be redundant and full of explanatory comments!

4.2 On Describing Data: Tables

After the data has been prepared in a meaningful way, one should familiarize oneself with the data in a next step. The aim of such a data description is always to get an overview of the distribution and certain properties of a variable. Description means usually to ignore certain information—descriptive statistics is an attempt to reduce complexity, to use once a fashionable term of sociology.

As a first step, the distributions of certain variables, such as gender role orientation or the underlying variables, should be presented. The possibilities of presentation differ according to the respective scale level of the variables. Normally, nominal, ordinal and metric variables are to be distinguished, for which—to repeat this basis of measurement once again—equality or difference, an order with respect to a specific dimension or also distances or ratios can be meaningfully interpreted. As nominal variables, we will use the example of religious affiliation and the question of whether one spent one's youth predominantly in the GDR or the new federal states or not, i.e. the place of socialization, in the following. The answers to the individual items on gender role orientation presented above are understood as ordinal data, data on the age of the respondents or also the index on the family image presented above as metric variables.

Table 4.1 Distribution of the items capturing gender role orientation (column percentages). (Source: GGSS 2016)

	Item 1	Item 2	Item 3
	Career	Division of labor	Housewife
Strongly disagree	38.7	41.8	48.6
Somewhat disagree	44.4	36.2	31.7
Agree	11.5	15.1	13.1
Strongly agree	5.4	7.0	6.6
n	1733	1741	1723

The simplest and certainly also best means of first describing data is the use of frequency tables. It cannot be emphasized strongly enough that the frequency tables generated by the corresponding data analysis programs should not be found in texts or publications of any kind in the form they are used there. They contain too much and too little information at the same time. Depending on the question of interest, the information is to be selected and presented. This selection has to be made by oneself and cannot be taken over by the data analysis program.

Here, in a first step, the three variables presented above, which together reflect the gender role orientations or the traditional family image, are to be examined in the GGSSS 2016. Here, the first interest is the percentage distribution of the different answer options. In addition to a meaningful title, the table should also contain information on the data source and the number of cases included in the analysis.[10] Table 4.1 shows the example just mentioned.

It is not possible to go into detail here about the possible errors and misunderstandings in the presentation of results, especially since there is an excellent presentation by Freeman et al. (2008) in this area, which mainly addresses possible errors.

"One of the easiest ways to display data badly is to display as little information as possible. This includes not labelling axes and titles adequately, and not giving units. In addition, information that is displayed can be obscured by including unnecessary and distracting details" (Freeman et al. 2008, p. 9).

---

[10] Due to missing values, these figures differ for the three items. In many applications, it is preferable to consider only those cases in the descriptive data that will also be included in the more complex models later. Since these items were part of the internationally designed so-called ISSP program, they were only presented to a part of the original GGSS sample. The number of cases here is therefore significantly lower than in other analyses.

As in many areas, less is usually more here, clarity of presentation is the highest premise and a simple guideline could be: Save ink (see again Freeman et al. 2008). For example, in the concrete case it is unnecessary to separate each line with lines, to give row percentages or the cell occupancies of the individual answers, since only differences and thus the (column-) percentages are of interest. Tables do not always have to take up the entire page width. In addition, the use of too precise information should ultimately be avoided. Are we really interested in differences in the fourth decimal place and do we trust our measurements so much that we can assume this accuracy? One should design the corresponding tables sparingly, even minimally. The clear recognizability of the answer options and the items is important.[11]

While the representation of variables measured nominally or ordinally with relatively few values is rather self-explanatory, problems can arise with ordinal variables with rather many values or metric variables. Above, a mean value index was formed from the items career, division of labor and housewife, which reflects a conservative family image. The number of possible values is simply relatively large here for a simple frequency table. The actual goal of a simple and easy to grasp representation of the empirical distribution would probably not be achieved with this.[12] This problem, however, occurs mainly with metrically measured variables such as age. For this reason, summarizing classes must be formed. Here, some perhaps self-evident rules should be followed, which are nevertheless violated again and again in practice.

- The class boundaries must be non-overlapping. Each individual observation, usually each individual person, must be able to be assigned unambiguously to a certain class.
- The classes must follow each other without gaps, so that each individual observation can also be assigned to a certain class. No observation should be left behind.
- The class widths should be as equal as possible, because only then can the descriptive results be meaningfully interpreted.

---

[11] These relatively simple rules are sometimes not followed by textbook publishers. For example, since the second edition of this book, we have been violating the rules and guidelines we formulated ourselves. However, this does not change the fact that they make sense and should actually be followed. The story of Cato the Elder shows that persistence can also lead to success.

[12] In principle, such a problem can also occur with nominal variables, for example, if the occupational activity was asked. In such cases, a sensible formation of occupational groups is required.

Table 4.2 Age distribution (column percentages). (Source: GGSS 2016)

Age group	Share (in percent)
Up to 30 years	14.1
30 to under 40 years	14.0
40 to under 50 years	17.1
50 to under 60 years	20.6
60 to under 70 years	17.4
70 to under 80 years	12.3
80 years and older	4.6
n	3486

In our example, the age of the surveyed person in the GGSS 2016, the range of values extends from 18 to 97 years. We therefore formed a first class that covers all persons under 30 years of age and then proceeded in 10-year steps and finally summarized all persons who are 80 years or older in a last group. At first glance, it becomes clear that we have not fulfilled the last condition, the same class width, everywhere. The first and the last age class are wider. However, if one changes this for the lowest class by forming classes for the persons from 18 to under 28 and then in further 10-year steps, one deviates from the common classification practice and also possibly creates confusion. In Table 4.2 the empirical distribution can be found.

To make it clear once again: Any kind of classification leads to a loss of information—but that is also the goal of descriptive analysis methods. In Table 4.2 we can, for example, not specify the proportion of people of a certain age. The aim is to reduce the abundance of information to a manageable level.

4.3   Distributions: Graphs and Measures

Besides presenting the data in tables, graphics can often help to get an impression of the distribution of the variables of interest. The variety of possible representations is almost unlimited, as a brief look at the respective options in the corresponding programs clearly shows—in some cases, however, these representations do little to help understand the distribution of the data. Often they confuse more or even create a false impression (see Freeman et al. 2008, pp. 9 ff.; Schnell 1994, pp. 4 ff.). This criticism applies, for example, to the three-dimensional representations or the almost ineradicable pie charts, cake or circle diagrams, which can be found again and again.

For perceptual reasons, for example, differences between different areas are relatively difficult to recognize—but this is the logic of circle diagrams. Since this is often also recognized by the authors, they then supplement the circle diagrams with the corresponding percentage values, so that the information is ultimately available twice. In practice, however, the entries are usually so small and low-contrast that none of the two information can really be recognized and read. A simple frequency table would then have been more sensible and space-saving and definitely preferable. Three-dimensional pie charts combine the outlined shortcomings in an elegant way.

In general, one should distinguish between analysis graphics and presentation graphics, because often data can be analyzed more easily and intuitively with the help of graphics (see for a multitude of examples still Schnell 1994). However, few of these graphics can be used in the context of presentations of the scientific results, especially in the interested public. One should always consider what purpose the creation of a graphic pursues: Does one as a researcher want to learn something about the data and possible relationships or does one want to present certain results in the public.[13]

If one takes seriously the perceptual-psychologically justifiable comprehensibility of different representations (see Schnell 1994, pp. 4 ff.) and has the objective that with the help of a graphic the distribution of a certain variable should be sensibly understandable, a comparison of the respective group memberships based on different lengths of bars and thus a bar chart seems almost inevitable. These bar charts serve to display certain characteristics for different values of a discrete variable—such as the relative frequencies of the different denominations. In Fig. 4.1 this diagram can be found for the data of the GGSS 2016. For reasons of better reproducibility, we have also refrained from weighting the data set here. However, to ensure that the planned overrepresentation of the new federal states in the GGSS does not distort the results, we have shown the distribution of denominations separately for East and West Germany.

This form of graphical representation is very suitable for nominal or ordinal variables with a manageable number of values. The visualization of the distribution of ordinal variables with a large number of values and of metric variables becomes more problematic.

---

[13] Many of these analysis graphics, such as probability plots or graphics in the context of cluster analysis, cannot be discussed here at all (see Schnell 1994 for this). In the context of residual analysis within multiple regressions, we will discuss some further methods below.

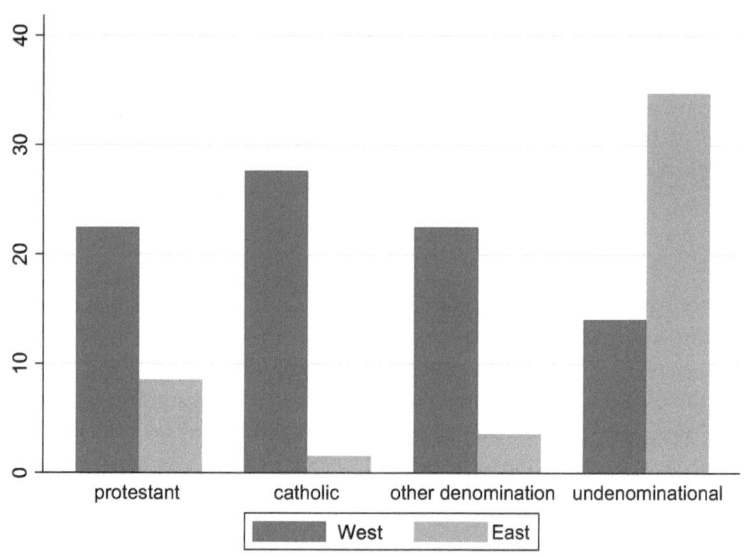

Fig. 4.1 Distribution of denominations by youth in East and West Germany (in percent). (Source: GGSS 2016, $n=3{,}183$)

Here, histograms are much more frequently used (see for further references Schnell 1994, pp. 21 ff.). Histograms also show the frequency distribution of classified data, but the area of each bar corresponds to the frequency in a class and thus ultimately frequency densities are displayed. With the help of these considerations, it is possible to construct and interpret histograms with different interval widths in a meaningful way. However, it is often not or at least not without great effort possible to implement this idea in practice. Here, only the width and the number of classes are variable, but not different class widths in one illustration. For this reason, histograms can usually be interpreted simply by comparing the height of the corresponding bars. Depending on the class width and thus also depending on the number of classes used, however, quite different impressions can be conveyed. It seems again sensible to generate histograms with different class widths to assess their differences (see for a discussion of the number of classes to be used sensibly Schnell 1994, pp. 21 ff.). In Fig. 4.2 there is a histogram of the age distribution, where we have chosen a class width of 10 and a starting point of 18.

One problem with these histograms is that for the individual density estimates only the persons in the corresponding range—in the example, in the lowest group

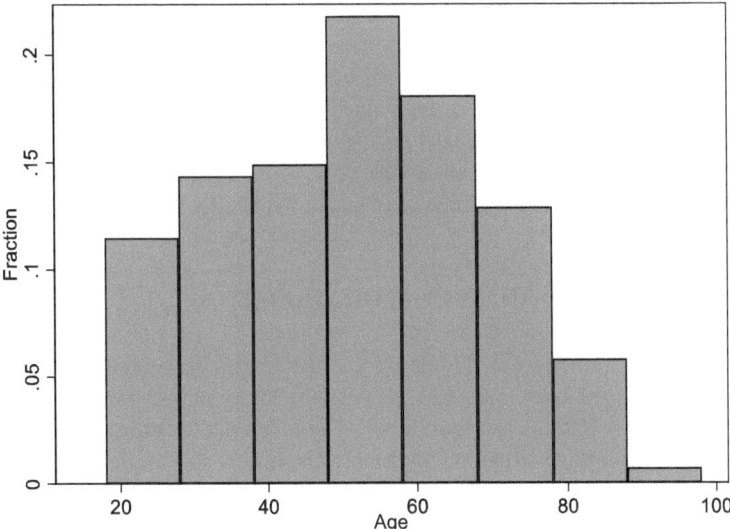

Fig. 4.2 Histogram of age. (Source: GGSS 2016, $n = 3,486$)

the persons between 18 and under 28—are used and this estimate does not vary in this group. One can now argue, however, that for the probability estimate, the direct neighborhoods—for example, for 27-year-old persons rather the 29-year-old people and not the 18-year-olds—should be used. This is exactly the logic followed by so-called kernel density estimators, which can also be easily generated in corresponding data analysis programs.

It is not possible to go into the multitude of other forms of presentation such as stacked and grouped bars, cylinders or even pyramids at this point—but this is not necessary either. One should always be clear about why one chooses a graphical representation and what one actually wants to say. The goal should be to make the understanding of corresponding distributions easier. Anyone who has ever tried to understand such—and we deliberately refrain from pointing out particularly unsuccessful representations—illustrations in detail will certainly choose other forms of presentation in the future.[14] However, if one wants to present the

---

[14] Finally, at least so-called stem-and-leaf plots for metric variables (see again Schnell 1994) must be mentioned. With the help of this method, one can quickly and very clearly present even larger amounts of data. Since stem-and-leaf plots are usually visually unappealing, they are unfortunately hardly used in the social sciences.

distribution of two metrically measured variables in a diagram, one should use a corresponding or scatterplot. However, since the number of realized attribute values is limited in most cases even for metric variables, these scatter plots are often not very informative, since individual attribute combinations are frequently occupied, but this is not recognized in simple scatter plots. To avoid this problem, there are various options, the easiest being to add a small random error to both variables, so that concrete attribute combinations are easier to recognize.

4.4   Measures: All for One, One for All?

Often there is an interest in describing the distribution of certain variables in a more concentrated form than can be done by frequency distributions or the graphical representations just introduced. A first, but not the only idea here is to summarize the entire distribution in a single number, as a measure of central tendency. Of course, such a simplification almost always involves a relatively strong loss of information—but that is, to repeat it, the basic idea. It always has to be taken into account how large the loss of information is and thus implicitly what error one accepts or wants to accept by choosing a certain measure of central tendency. Depending on the level of measurement of the respective variable, very different measures are available. For example, if one considers the distribution of religious affiliation and thus a nominal variable, one can determine as the indication with the highest information content the mode: the characteristic value that the relatively most people have. In our case, the GGSS data for 2016, the mode in the entire Federal Republic is the value "no denomination". If one had to guess the denomination of a randomly determined person without additional information, one would have the relatively best chances with the tip "no denomination", all other tips make more errors 'in the long run'.

The assessments of how informative such a measure is can understandably vary widely. They only gain real significance when one compares interesting subgroups and their measures of central tendency with each other. In the present case, this results in different modes for East and West Germany (see Fig. 4.1): While it is "no denomination" in the East, the indication of more than three quarters of all respondents, the value "Catholic" in the West—albeit narrowly—forms the most frequent case. When presenting empirical results, it is almost always useful to calculate various measures—both of central tendency and of the measures of dispersion and shape of distributions to be introduced below—and to present the information that is interesting and important for theoretical reasons afterwards.

If one now considers ordinal measured variables—such as the gender role orientations—one can of course also determine the mode here—it is at the first item at the value "tend to disagree" and at the other two at "completely disagree" (see Table 4.1). If one stops here, however, one disregards that the answer options contain a certain rank order, here an increasing agreement with a rather traditional family image. To take this property of the measurement into account, one uses the median of a distribution. The median is the value that divides an ordered distribution into two equal halves. For all three items, this is the value 2. The cumulative distribution exceeds the 50% mark at the value "tend to disagree" and thus this value represents the median. Of course, other quantiles can also be formed. Besides the median as the 50% quantile, quartiles and quintiles are common, but any conceivable percentiles can be easily generated.

The most well-known measure of central tendency is the arithmetic mean. Here, the metrically measured characteristics of all persons are added up and divided by their number. For n persons, this means formally:

$$\bar{x} = \frac{1}{n} \cdot \sum_{i=1}^{n} x_i$$

If one calculates this arithmetic mean, often simply called the mean, for example for the index of gender role orientation discussed above, it is 1.83 for all respondents on a scale of 1 to 4. The arithmetic mean minimizes the error defined as the sum of the squared deviations of the individual measured values from this measure and represents an optimal measure in terms of this definition.

4.5   Measures of Dispersion: "Birds of a Feather Flock Together"?

The measures of central tendency can now come about in very different ways. For example, for nominal variables, all groups can be almost equally populated or the overwhelming majority of observations have the same value. Similarly, attitude averages can result from either very many respondents having a medium value or there are two almost equally large groups at the extreme values of the scale. To capture this possible variability, there are several measures of dispersion.

Most of the known measures of dispersion assume a metric level of measurement. However, there are also some measures that can be used for so-called qualitative, i.e. nominal and ordinal variables. The basic idea is always that the measures take the value 0 when all persons have the same value, and the measure reaches or approaches the value 1 when the individual groups are of equal size. For this purpose, there are a number of proposals such as the deviance (Kühnel and Krebs 2001, pp. 96 ff.) or some entropy measures. A simple calculable measure is the index of qualitative variation (IQV) (Gehring and Weins 2009, pp. 130 f.). If pk is the relative frequency of the k-th of m possible classes of a distribution, the index is calculated according to the following equation:

$$IQV = \frac{1 - \sum_{k=1}^{m} p_k^2}{\frac{1}{m} \cdot (m - 1)}$$

If we now calculate separately for the above distribution of religious affiliation in the old and new federal states (cf. Fig. 4.1) the index of qualitative variation, we obtain a value of 0.52 for the east and 0.90 for the west. In the old federal states, the individual denominations are thus more evenly populated, while in the new federal states, the strong occupancy of the category "nodenomination" causes a relatively low index of qualitative variation.

However, much more common are the variance s^2 and the standard deviation s as measures of variation for metric variables. For the variance, the sum of the squared deviations of the values x_i on the metrically measured variable from the mean is formed for all persons and weighted with the number of cases n. Since the metric of this measure does not correspond to any intuitive interpretation, the standard deviation is often also calculated as the square root of the variance, which in turn corresponds to the original variable in its metric. Formally, the standard deviation is determined as follows:

$$s^2 = \frac{1}{n} \cdot \sum_{i=1}^{n} (x_i - \bar{x})^2$$

The variance is then simply $s = \sqrt{s^2}$. If we calculate these values for the above scale of gender role orientation, again separately for growing up in the GDR or the new federal states or the old Federal Republic, we can see that the values in the old Federal Republic show a much greater dispersion.

Another possibility to capture the variation of a variable is to use the percentiles presented above. Here, the distance between the first quartile, i.e. the 25%-quantile, and the third quartile (75%) is determined. As this description makes clear, this so-called interquartile range is only meaningful for metric variables, since only there the distance between two measurement points is interpretable.[15] Occasionally, the interquartile range is mentioned as a measure for ordinal data, but this only makes sense for rank orders.

More important is the interquartile range for a representation of distributions that has become increasingly popular in recent years for good reason: the so-called boxplot. A boxplot makes it very quick to obtain information about the center, symmetry, dispersion, skewness, and number and location of extreme values. A boxplot is defined by a box whose lower end is determined by the first quartile and whose upper end is determined by the third quartile. In addition, a line within the box marks the median. The length of the box thus corresponds to the interquartile range. In addition, the so-called 'whiskers' are drawn in a boxplot: "The upper or lower boundaries of the whiskers of a boxplot indicate the location of the 'inner fences'" (Schnell 1994, p. 19, translated by authors). The whiskers extend to the minimum or maximum values, but their length is at most 1.5 times the interquartile range. Values beyond the whiskers are shown individually, with a distinction between outliers, which lie at most three times the interquartile range outside the box, and extreme outliers or extreme values.

Of course, differences with respect to possible explanatory variables are also important here. For this reason, for example, in Fig. 4.3 a separate boxplot is created for each person who grew up in the GDR or the new federal states or not.

At first glance, Fig. 4.3 shows the clear difference in the distribution between East and West: Although the median is the same for both groups, the dispersion is much larger for those with a socialization in the old federal states. Those who grew up in their youth in the GDR or the new federal states rarely hold stronger traditional gender roles. Boxplots thus allow distributions to be captured very quickly.

4.6   Measures of Association

Already in these first descriptive presentations, but also in the introductory discussion on the objective of empirical social research, it became clear that pure description is only a first, albeit necessary, step. Theoretically much more

---

[15] In addition to the measures of central tendency, which are also called first moments, and the measures of dispersion, which represent second moments, there are also indicators that capture the skewness (skewness) as third moments or the kurtosis (kurtosis) as fourth moments. These results can usually be easily generated, but are very rarely actually interpreted.

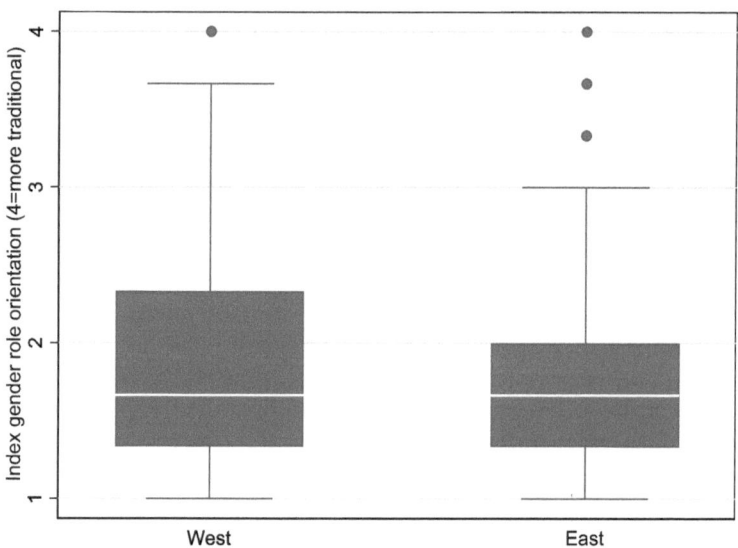

Fig. 4.3 Gender role orientation depending on the place of upbringing (East versus West). (Source: GGSS 2016)

interesting and exciting is the search for relationships between variables and thus—with all limitations and necessary specifications—also the search for causal influence processes. Before one can turn to these questions, one must examine whether there is a relationship between two variables at all.[16] Even if, for example, one can observe differences between the gender role orientations depending on growing up in East or West Germany, this does not necessarily have to be an effect that can be interpreted in terms of content, but can also be due to random fluctuations. In the following, various methods will be outlined that answer this question—true effect or random fluctuation—and then in a second step also give an answer to the question of how large the discovered relationship actually is. With regard to these measures, it is useful to think of some criteria in advance

---

[16] In Chap. 7 it will be discussed in detail that in the social sciences situations can quite frequently occur in which an actually existing relationship between two variables is concealed or suppressed by the influence of third variables. For the moment, we want to ignore this possibility for the further discussion.

that these measures should sensibly fulfil, because in order to be able to compare relationships, the measures should follow a uniform logic and metric (cf. still Benninghaus 1982):

- If there is no relationship between the two variables under consideration, a corresponding measure should take the value 0.
- With increasing relationship, the height of the measure should increase.
- In case of a perfect relationship, the measure should take the value 1—but not exceed it.

If one can distinguish the direction of the relationship, i.e. if the variables are ordinal or metric, the range of variation of the measure should lie between 1 for a perfectly positive relationship and −1 for a perfectly negative relationship.

Relationships between nominal variables If one now first considers the relationships between two nominal variables, one can, for example, analyse again the distribution of religious affiliation and socialisation in East or West Germany. In Table 4.3 the values already shown in Fig. 4.1 are given, but here only the absolute case numbers are displayed.

Here, clear differences can already be seen, whereby one can be almost certain, based on the differences, that the differences did not arise solely due to random processes. However, to statistically confirm this strong impression, one must compare this empirical distribution with the values that one would expect from a so-called indifference table, in which only the marginal distributions determine the cell frequencies. But what exactly does one mean by an indifference table?

In such a table, one would expect, for example, that the proportion of people with Catholic denomination is 30.4%—that is 969 divided by 3183—both with a West German and an East German socialization. Likewise, the proportion

Table 4.3 Religious affiliation and socialisation in West or East Germany (case numbers). (Source: GGSS 2016)

	West	East	Total
Protestant	742	277	969
Catholic	767	41	808
Other denomination	91	16	107
Non-denominational	425	874	1299
n	2025	1158	3183

Table 4.4 Denominational affiliation and socialization in West or East Germany (expected case numbers in case of independence). (Source: GGSS 2016)

	West	East	Total
Protestant	616.5	352.5	969
Catholic	514.0	294	808
Other denomination	68.1	38.9	107
Non-denominational	826.4	472.6	1299
n	2025	1158	3183

of people who grew up in the GDR or in the new federal states is 36.4% in all denominations. The Table 4.4 contains the corresponding expected values.

The difference between this expected distribution under independence and the real empirical distribution can now be summarized in a measure that is called χ^2—pronounced chi-square. If one denotes the observed count of the ij-cell in an $r \cdot c$-sized contingency table with $f_{b_{ij}}$ and the expected values under independence with $f_{e_{ij}}$, χ^2 is determined as follows:

$$
\chi^2 = \sum_{\substack{i = 1 \\ j = 1}}^{\substack{i = r \\ j = c}} \frac{(f_{b_{ij}} - f_{e_{ij}})^2}{f_{e_{ij}}}
$$

So, the squared deviations of the observed and the expected cell counts are formed, weighted by the expected value—since the same deviations are certainly more serious for smaller cell counts—and finally summed up. Since almost always deviations between the observed and the expected values occur, even if only because the expected values often have a decimal place, it must now be checked whether this deviation is due solely to random processes or to substantive relationships between the two variables. For this purpose, the empirical χ^2-value is compared with a corresponding probability density distribution. A test of the null hypothesis, there is no relationship, is used (see for the logic of these tests in general the Chap. 5). One looks at the probability that a corresponding χ^2-value occurs when there is no relationship between the data. If this probability is smaller than a predefined level, usually 5, 1 or 0.1%, the null hypothesis is rejected. In our case, the empirical χ^2-value is 969.6.

The probability density functions differ in their course for the number of degrees of freedom, which can be determined as $(c-1)\cdot(r-1)$ in a $c\cdot r$-sized table. For a $4\cdot 2$ table, there are therefore three degrees of freedom. With three degrees of freedom, however, the probability of obtaining a χ^2 value of 969.6 in a random table is well below 0.01%. For this reason, we assume that—as already evident from the distribution—the distribution between the individual denominations and the persons who grew up in East or West Germany is statistically significantly related.[17]

Now we know that there is a relationship between the two variables considered here, but we do not yet know how strong this relationship is. χ^2 values and their corresponding interpretation are sensibly determined by the number of cases. With relatively large samples, even small differences become significant—but this does not mean that they are always particularly important in terms of content. For this reason, as a second step, it was already specified at the beginning of this section to consider appropriate measures that capture the strength of the relationship and vary between 0 and 1 for nominal variables or between -1 and 1 for ordinal or metric variables. From the multitude of possible indicators, only Cramer's V is mentioned here (for further measures and their respective advantages and especially disadvantages, see Benninghaus 1982). Cramer's V is defined as follows:

$$V = \sqrt{\frac{\chi^2}{n\cdot\min[(r-1);(c-1)]}}$$

If r is the number of rows of a cross table and c is the number of columns, then the χ^2 value is divided by the number of cases, multiplied by the smaller value of $(r-1)$ and $(c-1)$, and the root of this quotient is formed. In our case, Cramer's V is determined as follows:

$$V = \sqrt{\frac{696.6}{3183\cdot 1}} = 0.55$$

---

[17] Generally, a warning must also be given at this point against a theory-free approach to data analysis. If one simply examines the data for significant relationships in the sense of a crude empiricism, one must be aware of the fact that even in pure random data 5 out of 100 tests show a significant relationship—this is the logic of these tests.

Since Cramer's V can only vary between 0 and 1, the relationship between the denomination and the place of socialization in East or West Germany is relatively strong.[18]

A different logic follows the measure λ—pronounced lambda. It follows the so-called PRE logic, where PRE stands for "proportional reduction in error". All PRE measures implicitly assume a distinction between an independent and causally prior and a dependent, influenced variable. PRE measures then examine to what extent a first estimate of the dependent variable can be improved by using the values of the independent variable for the prediction. In a first step, for example, the distribution of religious affiliation in Germany is considered without taking into account whether a person grew up in East or West Germany.

If one had to guess the religious denomination of a certain person, the best strategy would be to bet on the mode of the dependent variable—here, on the attribute "no denomination". In 40.8% of all cases, this would result in a correct prediction—but in 59.2% or 1884 cases, it would also make a mistake, the type I error E1. To what extent can this error be reduced if one knows the value of the variable East-West and takes it into account in the prediction? For this, the Table 4.4 has to be looked at again.

If one follows the same logic here, one would now assume the prognosis "Catholic" for the West, but remain at the value "no denomination" for the East. Of course, one also makes mistakes here—namely with all members of other religious communities and non-denominational people in the West, as well as those of the Catholic faith, Protestants and members of other religious communities in the East. In total, this concerns 1542 people. This is called the type II error or E2. λ is now calculated as $(E1 - E2)/E1$, i.e. as $(1884 - 1542)/1884$ and is thus 0.18. 18% of the errors in predicting the religious affiliation can thus be avoided if one knows whether the person grew up in Eastern or Western Germany.

Relationships between ordinal variables An important goal of social science research is to collect the most diverse theoretical constructs and variables of interest as well as possible and that means as informatively as possible. This means

---

[18] Occasionally, one finds in textbooks classifications of when individual statistical measures can be considered as low or weak, medium or strong. Since social science models usually always leave out and often even have to leave out a multitude of explanatory factors, this tradition should not be followed, as these classifications always have to be arbitrary. Of course, one still has to think about the extent to which the statements have any content relevance at all.

Table 4.5 Relationship between two ordinal variables. (Source: GGSS 2016)

Woman should help man with career	Man in professional life, woman household and children			
	− −	−	+	++
Strongly disagree	442	168	44	15
Somewhat disagree	235	356	136	41
Agree somewhat	32	73	59	36
Agree completely	18	28	19	28

that one usually strives to achieve the highest possible level of measurement. In data analysis, this additional information can then be used. For example, if it is possible to capture not only the equality or inequality of a certain variable and thus measure it nominally, but at least a certain order of the answers, this information should be taken into account in data analysis. In empirical social research, ordinal information is very often available for individual items—we discussed this above in the context of the items on gender role orientation.

Even though the measures just mentioned can also be determined for the analysis of the relationship between ordinal variables, this would mean giving up the additional information. In this section, a few indicators will be introduced that take into account this fact of the ordered answers. The representation of two ordinal variables is usually done again by a contingency table. In Table 4.5 the distribution of the respondents according to the first two items presented above, which we called "career" and "division of labor", can be found. To simplify the presentation, the rejecting categories in the header of the table were abbreviated with − − or − and the positive categories with + and ++.

The basic logic of all measures for ordinal relationships is based on the comparison of all possible pairs that can be formed from the 1730 respondents. In total, almost 1.5 million of these unordered pairs—exactly 1,495,585 pairs—have to be considered. For each of these pairs, one of the following five results can be observed:

- The person A agrees more with the first item than the person B and the same applies to the second item. The persons A and B then form a concordant pair. Conversely, it follows that the person B has a more negative attitude towards both items than the person A. The number of these pairs can be denoted by C.
- The person A can again rate the first item more positively than the person B, who, however, rates the second item more positively than the person A. Then it is a discordant pair, the number of these pairs is denoted by D.

- In addition, it is possible that the two persons agree on the assessment of the first item, but have a different opinion on the second item. Then it is a pair tied in item 1, a so-called tie. Their number is T_{Item1}.
- An equivalent constellation occurs when the two persons A and B agree on the second item, but have a different opinion on the first item. T_{Item2} is the number of these pairs.
- Finally, it is still possible that the two persons rate both items equally. The number of these pairs is then finally $T_{\text{Item1;Item2}}$.

A multitude of measures can now be determined, which ultimately all consider the difference between the concordant and the discordant pairs—i.e. the first two constellations presented. The differences between the various indicators then lie in the ratio in which this difference is considered. The simplest measure is Goodman and Kruskal's γ, pronounced gamma, which is determined as follows:

$$\gamma = \frac{C - D}{C + D}$$

If only concordant pairs are observed, γ takes the value 1, if there are only discordant pairs the value -1. In our example, the corresponding value is 0.56—the two attitudes are thus relatively highly correlated. Now contingency tables can be determined, in which, for example, γ takes a high value, although one has the feeling that ultimately there is no relationship between the variables, for example because almost all persons agree on one item and only very few pairs are rated as concordant. For this reason, some further measures have been developed, which are characterized by refinements in the denominator. For example, Kendall's τ_b—pronounced tau b—is determined as follows:

$$\tau_b = \frac{C - D}{\sqrt{(C + D + T_{\text{Item1}}) \cdot (C + D + T_{\text{Item2}})}}$$

In practice, γ is most often reported, as it always gives the highest value of all measures of association for ordinal variables. These measures are especially important because they can take into account the additional information of ordinal variables with the logic of pairwise comparisons. A reference, for example, to the distance from the median is not meaningful, as the distances are not interpretable for ordinal measurements. Despite the charm of these different measures for ordinal variables, one has to admit at the end of this section that all the measures discussed here and anyway presented rather selectively have their disadvantages and are rarely used in practice. More common is the so-called rank correlation coefficient according to Spearman. Here, the persons are listed according to their ranking on two variables. If d_i is the difference of these rank positions with respect to the assessment of person i for the two items, the rank correlation coefficient is determined as follows:

$$r_s = 1 - \frac{6 \cdot \sum_{i=1}^{n} d_i^2}{n \cdot (n^2 - 1)}$$

Relationships between nominal and metric variables Many theoretically interesting quantities, however, are assigned a metric level of measurement. This means that the differences between two measurements can be meaningfully interpreted. For example, a scale of gender role orientation was presented above. For metric measurements, one can then determine the arithmetic mean as a measure of central tendency. An interesting question then is, for example, whether the average attitudes differ between East and West Germany. As a graphical form of presentation, grouped boxplots can be used (see Fig. 4.3).

Even if the difference in this question is relatively clear, two questions must now be differentiated: Is firstly the mean difference between East and West Germany substantively significant or can it perhaps be attributed to random fluctuations? And if there is an true effect: How large is then the second relationship between gender role orientation and growing up in East or West?

To answer the first question, so-called t-tests are usually calculated. For this, the difference of the group means is weighted by their standard error.

$$t = \frac{\bar{x}_{\text{Group1}} - \bar{x}_{\text{Group2}}}{\text{Standard error}_{\bar{x}_{\text{Group1}} - \bar{x}_{\text{Group2}}}}$$

The standard error is determined by the standard deviation of the index and the respective size of the two groups.[19] The test statistic follows the so-called t-distribution. As a rule of thumb, it can be said that from a value of about 2, one can assume that the groups differ. Here, too, the analysis programs usually give the exact results.

However, the question of whether this relationship can again be summarized in a single measure is much more exciting. For this, one can again use the PRE logic discussed in detail above. It is thus examined to what extent the knowledge of the value of the independent variable, i.e. the question of whether the persons spent their youth in East or West Germany, leads to an improvement of the assessment

---

[19] In detail, it must still be checked whether one can really assume for both groups that the variance and thus the standard deviation, which is necessary for the calculation of the standard error, can be assumed to be identical. There are also statistical tests for this.

of the gender role orientation. The error is understood as the squared deviation of the predicted from the actual value of a person. Thus, it holds:

$$\text{Error}_{\text{Person}_i} = (\hat{y}_i - y_i)^2$$

In a second step, it must be determined which method is used for the prediction of the respective characteristic value. It can be shown that under the above error definition, the mean gives the best predictions. What error is made if one assumes the mean of the index for all respondents, i.e. regardless of the place of their socialization?

$$\text{Error}_{\text{Type1}} = \sum_{i=1}^{n} (y_i - \bar{y})^2$$

To what extent does the prediction improve if one is informed whether a respondent was socialized in East or West Germany? Well: For this, one considers the error that is still made if one now uses the respective mean for the two groups as the prediction value instead of the mean of all respondents.

$$\text{Error}_{\text{Type2}} = \sum_{k=1}^{2} \sum_{i=1}^{n_k} (y_i - \bar{y}_k)^2$$

The error of a person is now determined as the squared deviation of his or her value from the respective group mean. It can now be shown that the difference between the error of the first type and the error of the second type can be understood as the explanatory power of the group distinction and that consequently it holds:

$$\sum_{i=1}^{n} (y_i - \bar{y})^2 - \sum_{k=1}^{2} \sum_{i=1}^{n_k} (y_i - \bar{y}_k)^2 = \sum_{k=1}^{2} \sum_{i=1}^{n_k} (\bar{y} - \bar{y}_k)^2$$

In the case of the above example, there is only a relatively small difference between the two groups and the within-group variance far outweighs the group differences. If one now forms the ratio of the avoided errors, the improvement of the prediction, to the originally committed errors, one obtains the size η^2—pronounced eta-squared. In our case, η^2 is just 0.009, not even one percent of the errors can be avoided.

Relationships between metric variables Finally, at the end of this chapter, we will address the relationship between two metrically measured variables. Since we can now use the arithmetic mean as a measure of central tendency, we do not have to perform all possible pairwise comparisons as we did for the measures of association for ordinal variables, but we can look at the deviations of the values

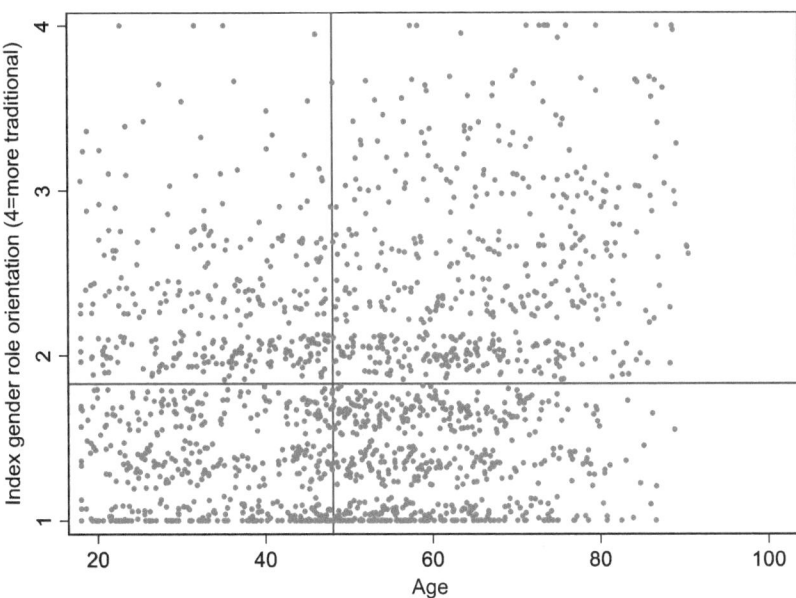

Fig. 4.4 Scatter plot between gender role orientation and age. (Source: GGSS 2016)

of the two variables of interest for a person from the respective mean value. In the following, we will look at the relationship between the now well-known index of gender role orientation and the age of the respondents. We take a look at the position of the individual cases in relation to the mean values. Here we have four cases, which can be traced by using Fig. 4.4. The scatter plot is divided into four quadrants based on the arithmetic mean values of the two variables:

- If a person is younger than the average, they also have a more liberal gender role orientation. In the scatter plot, they are in the lower left quadrant.
- A person is older than the average and also has an above-average traditional gender role orientation. They are therefore in the upper right quadrant.
- A person is below average in age, but above average in traditionality (quadrant top left).
- And finally, a person can be above average in age, but below average in traditionality (quadrant bottom right).

In the first two cases, we have a concordant or positive deviation, in the last two cases, a discordant or negative deviation of a person from the respective mean values. In the scatter plot, it can already be guessed that there are more cases in the quadrants bottom left and top right than top left and bottom right. This indicates a positive relationship between the two variables: The older, the more traditional. Since a metric scale level is present, we can combine the positive or negative deviations and simply sum them up. If we normalize this quantity by the sample size, we obtain the covariance of two variables:

$$\text{Covariance}_{x,y} = \frac{1}{n} \cdot \sum_{i=1}^{n} (x_i - \bar{x}) \cdot (y_i - \bar{y})$$

Since this quantity depends on the exact measurement and the metric of the individual variables, it is weighted by the root of the product of the variances, i.e. their geometric mean. The resulting quantity is called the correlation coefficient, or more precisely: the Pearson product-moment correlation coefficient.

$$r_{x,y} = \frac{\frac{1}{n} \cdot \sum_{i=1}^{n} (x_i - \bar{x}) \cdot (y_i - \bar{y})}{\sqrt{(\frac{1}{n} \cdot \sum_{i=1}^{n} (x_i - \bar{x})^2) \cdot (\frac{1}{n} \cdot \sum_{i=1}^{n} (y_i - \bar{y})^2)}}$$

As dramatic as this formula may always look, it ultimately captures simply the same- and opposite-direction deviations from the respective mean. However, by means of the corresponding standardizations, here the denominator of the formula, one achieves a number of very desirable properties.

- Correlations are symmetric.
- Correlations lie between -1 and 1.
- Positive values indicate a positive relationship. There is therefore a 'the more, the more' relationship.
- Negative values indicate a negative relationship and thus a 'the more, the less' relationship.

In the example, age and the index of gender role orientation correlate with 0.19.[20] It almost goes without saying that it can also be determined for the correlation coefficient whether it came about due to random processes.

---

[20] In more in-depth analyses with longer time series, it would now have to be checked whether this relationship can be differentiated into an age effect—respondents are more traditional because they have progressed further in their life course—and into a cohort effect—respondents are more traditional because they were socialized in certain temporal contexts—(see e.g. Lois 2020).

4.7 Afterword

This chapter dealt with the first steps of data analysis. First, the data preparation, which is certainly the most extensive area in practice, but mostly neglected in the literature, was discussed. In this field, a multitude of decisions are already made that influence the further analysis—usually unconsciously and in most cases hardly ever traceable in publications. In order to ensure clarity and reproducibility in this area, the use of syntax files and thus the documentation of one's own approach, but also the open handling of these files, is indispensable.

In a second step, the description of the data to be analyzed was addressed: Often, less is more in these analyses, especially in their presentation. One should be very clear about what one actually wants to say—the technical possibilities available today probably generate too many rather than too few options and often, for example, graphics are more confusing than enlightening and thus have missed their purpose.

The focus of this chapter, however, was the analysis of relationships between variables. Depending on the scale level of the variables under consideration, different possibilities arise. The most important measures of association are summarized again (Table 4.6):

Almost consistently, one can distinguish between two questions: Do certain characteristics arise solely due to random fluctuations or can certain substantive processes be assumed? How strong is the second correlation between the two variables considered, whereby the correlation measures for nominal variables should ideally vary between 0 and 1 and for ordinal and metric variables between -1 and 1. As further reading for the first steps of data analysis, we recommend Jones and Goldring (2022), who give a comprehensive introduction to exploratory and descriptive statistics and also cover the visualization of data. An introduction to data analysis with Stata is provided by Acock (2022). For the visualization of data using Stata we recommend Mitchell (2022).

Table 4.6 Measurement level and measures of association. (Own representation)

Independent variable	Dependent variable		
	Nominal	Ordinal	Metric
Nominal	χ^2;λ; Cramer's V		η^2
Ordinal		τ_b; γ; r_s	
Metric			r

References

Acock, Alan C. 2022. *A Gentle Introduction to Stata. Revised Sixth Edition.* College Station, Texas: Stata Press.

Benninghaus, Hans. 1982. *Deskriptive Statistik. Studienskripten zur Soziologie 22.* Stuttgart: Teubner. https://doi.org/10.1007/978-3-322-93052-1.

Brecht, Bertolt. 1977. *Gesammelte Werke 10. Gedichte 3.* Werkausgabe edition suhrkamp. Frankfurt a. M.: Suhrkamp.

Freeman, Jenny V., Stephen J. Walter, and Michael J. Campbell. 2008. *How to display data.* Malden: Blackwell.

Gehring, Uwe W., and Cornelia Weins. 2009. *Grundkurs Statistik für Politologen und Soziologen.* Wiesbaden: VS Verlag. https://doi.org/10.1007/978-3-531-91879-2.

Hahn, Alois, Johannes Kopp, and Nico Richter. 2019. „Zwei Freunde und doch so verschieden". Vorstellungen von Partnerschaft, Ehe und Familie in einer Beziehung: Ein Vergleich der Perspektive von Frauen und Männern. In *Erklärende Soziologie und soziale Praxis,* Eds. von Daniel Baron, Oliver Arránz Becker, and Daniel Lois, 215–250. Wiesbaden: Springer VS. https://doi.org/10.1007/978-3-658-23759-2_10.

Jones, Julie Scott, and John Goldring. 2022. *Exploratory and Descriptive Statistics.* London: SAGE Publications.

Konrath, Sara, Brian P. Meier, and Brad J. Bushman. 2014. Development and validation of the Single Item Narcissism Scale (SINS). *PLoS ONE* 9:1–15. https://doi.org/10.1371/journal.pone.0103469.

Kühnel, Steffen-M., and Dagmar Krebs. 2001. *Statistik für die Sozialwissenschaften. Grundlagen, Methoden, Anwendungen.* Reinbek: Rowohlt. https://doi.org/10.1007/s11615-003-0126-9.

Lois, Daniel. 2020. Gender role attitudes in Germany, 1982–2016: An age-period-cohort (APC) analysis. *Comparative Population Studies* 45:35–64. https://doi.org/10.12765/CPoS-2020-02.

Mitchell, Michael N. 2022. *A Visual Guide to Stata Graphics. Fourth Edition.* College Station, Texas: Stata Press.

Munsch, Christin L. 2015. Her support, his support: Money, masculinity, and marital infidelity. *American Sociological Review* 80:469–495. https://doi.org/10.1177/0003122415579989.

Munsch, Christin L. 2018. Correction: Her support, his support: Money, masculinity, and marital infidelity. *American Sociological Review* 83:833–838. https://doi.org/10.1177/0003122418780369.

Schnell, Rainer. 1994. *Graphisch gestützte Datenanalyse.* München: Oldenbourg. https://doi.org/10.1515/9783486787320.

Significance Test

The logic of significance testing is not necessarily easy to understand. However, once you have grasped the principle, you can apply this logic to a whole range of different inferential statistical tests—so it is worth spending some time on the topic of significance testing.

The aim of a significance test is to examine whether there is a relationship between two (or more) variables or not. For example, does a person's age affect whether they are more or less opposed to traditional gender roles? In other words: Are the two variables 'age' and 'traditional gender role orientation' in a statistically significant relationship?[1] We will explain the logic of significance testing right at the beginning and do so in a relatively abstract way. Later, we will also illustrate the significance test using examples.

The logic of significance testing can now be described as follows: We imagine a world in which there are no relationships between variables. For example, educational opportunities are independent from social background, just as motivation is not related to career success. For health, it is irrelevant how we eat and general life satisfaction is independent of the quality of our social relationships. In this world, traditional gender role orientation would also be independent of a person's age. We imagine this 'world without relationships' because we calculate certain test statistics for the significance test, whose distribution is known to us in

---

[1] Here, a causal hypothesis could also be formulated, since only the direction of effect from age to gender role orientation is possible. For the sake of completeness, it should also be noted here that a (causal) link between age and traditional gender role orientation in the case of a single measurement point could also be due to a cohort effect (see Lois 2020). For the definition of gender role orientation, see Krampen (1979) or Chap. 7 in this volume.

F. Hartmann et al., *Social Science Data Analysis*, https://doi.org/10.1007/978-3-658-41230-2_5

this world. This means that we know which values of the test statistic are likely under the assumption that no relationships exist, and which values are rather or extremely unlikely. In the world where no relationships exist, we are familiar and can estimate which values we can expect for our statistical test statistic.

If we now obtain a value for the test statistic based on our sample that is likely in the 'world without relationships', then we also assume that there is actually no relationship between the variables examined. In this case, we would assume that a person's age and their gender role orientation are independent of each other. However, the situation is different if we obtain a value for the test statistic that is extremely unlikely in the 'world without relationships'. In this case, we could still stick to the assumption that there is no relationship between the variables examined and assume that we were extremely lucky to come across such an unlikely value based on the sample data. But instead, we conclude by implication that the reason for this extreme value of the test statistic is that there is indeed a relationship between the variables examined. In this case, we would come to the conclusion that the variables age and traditional gender role orientation vary depending on each other. We will clarify later in the chapter what exactly is meant by 'unlikely' or 'extremely unlikely' and how we draw the boundary for this. We will also show how to calculate and interpret corresponding test statistics. We proceed as follows: Using two examples, we will each perform a hypothesis test step by step and explain each individual step in detail. Here it becomes clear that the principle of the significance test is universally applicable. Subsequently, however, it will also turn out that despite the same principle, different inference tests and respective test statistics are available and must be selected for one's own question according to certain criteria. Not every result of a hypothesis test is relevant from a social science perspective and the result of a hypothesis test should always be viewed with a healthy skepticism—in this context, we discuss what the terms effect size and statistical power mean. Explanations of the p-value, which is important for the interpretation of the outputs of statistical programs, as well as the relationship of hypothesis tests to confidence intervals, together with concluding remarks, form the end of the chapter. In the following three subchapters, however, we would like to deal in advance with some central terms and concepts for the significance test.

5.1   Basic Terms

The *significance test* is also called a hypothesis test, as it is checked whether a certain hypothesis should be retained or rejected. A *hypothesis* is an assumption about the population that relates to the relationship between two (or more) variables

(Hartmann and Lois 2015). In the significance test, we set up two types of hypotheses, the *null hypothesis* (H_0) and the *alternative hypothesis* (H_1). These are a pair of opposing hypotheses. This means that one hypothesis expresses exactly the opposite of the other in terms of content. The null hypothesis always assumes that there is no relationship between variables. The alternative hypothesis claims the opposite and states that there is a relationship in the population. The *population* is the set of all objects (in the social sciences these are usually persons) about which we want to make statements, and should be defined exactly before the data collection. In the German General Social Survey (GGSS), for example, the population consists of the adult resident population in Germany. Due to limited resources, we as social scientists do not survey the entire population, but only draw a subset from it, our *sample*. From the persons of the sample, we collect certain variables, which we suspect based on scientific theories to be connected in a certain way (see also Chap. 7). We therefore already have certain assumptions about the population before the data collection, which we can check afterwards with the help of the sample data by performing a significance test.

5.2 Statistical Hypotheses

With the help of so-called *inferential statistics* we are able to make statements about a population based on sample data. What we can learn about the population within the framework of inferential statistics can be roughly divided into two areas, estimation and testing (Ludwig-Mayerhofer et al. 2014). In *estimation* we try to estimate certain characteristics of the population, so-called population parameters, such as the population mean, by calculating a single value (point estimate; e.g. sample mean) or a range of values (interval estimate; e.g. 99% confidence interval; see also Sect. 5.10) based on our sample data. In *testing*, on the other hand, we formulate in advance the null and the corresponding alternative hypothesis regarding the relationship between the variables of interest and calculate a *test statistic* based on our sample data, on the basis of which we decide whether to retain the null hypothesis or reject it in favor of the alternative hypothesis. It should be mentioned at this point that although we decide for the null or the alternative hypothesis, we can never be sure whether we have made the right decision. There is always a residual risk of error. At least, however, we are able to indicate the probability of being wrong if we decide for the alternative hypothesis as a result of the calculated test statistic.

As already mentioned, we always formulate a pair of hypotheses, the null hypothesis and the corresponding alternative hypothesis, when performing a statistical test. With the null hypothesis, we assume that there is no relationship between the variables that we are interested in. With the alternative hypothesis, we assume the opposite, namely that the variables are related and vary depending on each other. Alternative hypotheses are usually derived from the scientific literature, based on theories and the results of previous studies (on the derivation and formulation of hypotheses, see Hartmann and Lois 2015). Depending on the state of research, we can formulate either directional or non-directional hypotheses. For example, if we suspect that the variables age and traditional gender role orientation are related, but cannot infer from the literature how exactly the variables are connected, we formulate a *non-directional* alternative hypothesis. If, on the other hand, we have reason to assume that the relationship should have a certain direction, then we formulate a *directional* hypothesis. For example, if we assume that traditional gender roles are more likely to be agreed with when age is high (or that a traditional gender role orientation is more likely to be rejected when age is high), then we already indicate this positive (or negative) direction in the formulation of the alternative hypothesis and are able to make statements with a higher information content than in the case of non-directional hypotheses.

As soon as we have formulated the hypotheses using linguistic means, we can transform them into *statistical hypotheses*. This usually involves relational operators (e.g. $<, >, =$) and Greek letters (e.g. μ, ρ). The latter symbolize the characteristics of the population. We want to illustrate this briefly with an example. In the following, we assume that after an extensive literature review we come to the conclusion that the metric variable age should be related to the variable traditional gender role orientation, which is assumed to be metric, and that we want to test the relationship using a significance test. In Sect. 4.7 we have already learned that the relationship between two metric variables can be quantified with the product-moment correlation coefficient r. The correlation coefficient r is calculated based on the sample data and indicates the relationship in the sample. However, in inferential statistics we are more interested in the population than in the sample. Thus, the sample in inferential statistics serves 'only' the purpose of gaining insights about the population. This means that we calculate r based on the sample, but we want to find out whether there is a relationship in the population and therefore formulate our hypotheses always with respect to the characteristics in the population. The symbol for a correlation in the population is ρ (small Greek letter rho). With the null hypothesis we assume that there is no relationship. For didactic reasons, we initially assume that we have not found any well-founded assumption in the literature about the direction of the relationship between age and gender role orientation. The statistical null hypothesis is:

$$H_0 : \rho = 0$$

The statistical alternative hypothesis is formulated in the undirected (two-sided) case as follows:

$$H_1 : \rho \neq 0$$

Two-sided means that both 'significantly' negative and 'significantly' positive deviations of the product-moment correlation from the value zero are an indication for the H_1. If we could infer from the literature that the relationship should be positive, we could set up a directed (one-sided) pair of hypotheses:

$$H_0 : \rho \leq 0 \quad H_1 : \rho > 0$$

And finally, the pair of hypotheses in the case of a presumed negative relationship would be:

$$H_0 : \rho \geq 0 \quad H_1 : \rho < 0$$

In this case, we would assume that there is a negative relationship in the population and that the traditional gender role orientation is less pronounced the older a person is. If there is a theoretically well-founded possibility to formulate a directed hypothesis, it should always be taken, as the gain in knowledge of the significance test is higher in the directed variant than in the undirected one.

5.3   Types of Errors and Significance Level

When testing hypotheses, we can never be completely sure that we have made the right decision in the end. We can make two errors. If we decide, based on our sample data, to reject the H_0 in favor of the H_1 and thus finally assume that there is a relationship in the population, we may commit the *type I error,* which is also called *alpha error.* If we draw our sample randomly from the population, it is therefore quite possible that our sample data (by chance) speak for the H_1, although there is actually no relationship between the variables of interest in the population. For our example, this would mean that we have randomly drawn a sample on whose basis a product-moment correlation coefficient is calculated that indicates a relationship between age and gender role orientation, although in the population there is actually no connection between the variables. But our sample data can also indicate the retention of the H_0. In this case, we may commit the *type II error,* also called *beta error.* As a result, we would conclude at the end of the hypothesis test that there is no relationship in the population, although the variables of interest are actually related in the population—we just don't know that. An overview of the types of errors is given in Table 5.1. Here it becomes

Table 5.1 Types of errors in significance testing. (Own representation)

		Decision based on the sample for	
		H_0	H_1
In the population, the following holds	H_0	Correct	Alpha error
	H_1	Beta error	Correct

apparent that we can also decide correctly. But no matter how we decide, there is always a residual risk that we are wrong. When testing hypotheses, the goal is usually to keep the probability of an alpha error low. Thus, we stick to the null hypothesis relatively long until the test statistic shows such an extreme value that we have to reject it in favor of the alternative hypothesis. For this purpose, the so-called *significance level* is determined before the significance test, which is denoted by the letter α. With α we indicate the probability of the alpha error (decision for H_1, although H_0 is correct) and thus determine which *error probability* we are willing to accept when deciding for the H_1. If we set α to 5%, for example, this means that we are willing to be wrong with a probability of 0.05 if we reject the H_0 as part of the significance test.

There are certain conventions for setting the significance level. Usually, α is set to 0.1% (common symbol: ***), 1 % (common symbol: **), 5% (common symbol: *) or even 10% (common symbol: #). Which significance level is appropriate is also a substantive decision. For example, the error in the social sciences may be less fatal than in medicine, where it can be life-threatening if the assumption formulated with H_1 regarding the effect of a drug does not correspond to reality. In such a case, a α of 0.1% or less would be preferable.

5.4   Step by Step using the Example of a Correlation Hypothesis

A hypothesis test can always be carried out according to the same scheme. We will present seven steps below that ultimately lead to a decision for or against the null hypothesis (cf. Clauß et al. 2017, p. 180; Hartmann and Lois 2015, p. 29).

1. Assumptions
2. Null and alternative hypothesis
3. Significance level
4. Test statistic

5. Critical region
6. Calculation of the test statistic based on the sample data
7. Decision for or against the null hypothesis

To apply the *seven-point scheme*, we refer back to our example. We want to test whether there is a correlation between the variables age and traditional gender role orientation (H_1) or not (H_0). For this, we use the data from the GGSS, more precisely the GGSS Cumulation 1980–2018, and here the data from the year 2016. To measure the traditional gender role orientation, we form the mean value of the variables fr02 *("It's more important for a wife to help her husband with his career than to pursueher own career")*, fr06 *("A married woman should not work if there are not enough jobs to go round and herhusband is also in a position to support the family")* and fr04a *("It is much better for everyone concerned if the man goes out to work and thewoman stays at home and looks after the house and children")*. The items could be answered by the respondents with values between 1 *("Completely agree")* and 4 *("Completely disagree")*. We recode the mean value of each person by subtracting it from the value 5, so that small values indicate a low agreement and large values indicate a strong agreement with traditional role models.

Assumptions First, the assumptions on which a test is based have to be clarified. This step is admittedly not easy at the beginning, as it already requires some more extensive knowledge about which test is suitable for which question and what prerequisites to pay attention to. We will also come back to this in Sect. 5.6. If the prerequisites for a test are not met, but the test is performed anyway, the results may not be reliable to interpret. For such a case, there are alternative non-parametric tests available. Also at this point, we refer to Sect. 5.6.

That the sample(s) of the study were drawn randomly is an important prerequisite for inferential statistical methods. Further prerequisites that have to be considered in the first step of the significance test concern, for example, the level of measurement of the variables of interest and the distribution of the variables in the population. If, as in our example, the linear relationship between two metric variables is to be determined and both variables are normally distributed, the product-moment correlation coefficient and a corresponding test statistic are usually determined to test whether a relationship can be assumed in the population or not. Prerequisites for this significance test are thus that the relationship between the variables is *linear*, that both variables are *metric* scaled and that both variables have a *normal distribution*. Strictly speaking, the respective normal distribution is not sufficient, but actually the joint distribution of the two variables must also follow a normal distribution. However, this is rarely tested. In some statistical programs, there is also no procedure available to check multivariate distributions. The respective normal distribution of the variables can, however, (as a necessary, but not sufficient condition) be checked by looking at corresponding graphs such as histograms or boxplots or by performing a significance test to check the normality assumption. Whether the relationship between the two variables is linear (or, for example, curvilinear) can be decided with the help of a so-called scatter plot. Here, the values of the two variables are displayed as points in the two-dimensional space. This can show whether the cloud of points has a straight-line course (see, for example, Fig. 6.1).

We assume in the following that our two variables are measured on a metric scale, have a linear relationship and are each normally distributed, and accordingly decide to perform a significance test based on the product-moment correlation coefficient.

Null and alternative hypothesis For didactic reasons, we first assume that the variables age and traditional gender role orientation have a linear relationship without specifying the direction and formulate our statistical hypotheses accordingly. First, the null hypothesis:

$$H_0 : \rho = 0$$

Then we formulate the undirected alternative hypothesis:

$$H_1 : \rho \neq 0$$

Significance level In this step, we determine the significance level, i.e. the probability of committing a type I error. Considering the sample size ($N = 1,742$), we choose a significance level of 1% ($\alpha = 0.01$) to account for the fact that even small and possibly irrelevant correlations become more likely to be significant with increasing sample size.

$$\alpha = 0.01$$

It should be emphasized at this point that we set the significance level in advance. It is not permissible to adjust the level after calculating the test statistic in order to report a significant result in the end.

Test statistic The test statistic for testing a linear relationship between two metric variables that are (bivariate) normally distributed in the population is calculated based on the product-moment correlation coefficient:

$$t = \frac{r_{x,y} \cdot \sqrt{n-2}}{\sqrt{1 - r_{x,y}^2}}$$

The advantage for us is that the distribution of this test statistic is known under a certain condition: the calculated t-values follow a t-distribution (see dotted or dashed line in Fig. 5.1) with the expected value zero, if the null hypothesis is true. The shape of the t-distribution depends on the so-called *degrees of freedom* (*df*). They indicate for a certain statistic how many values can vary freely for its determination (Bortz and Schuster 2010). The expression given above is t-distributed with $df = n - 2$ degrees of freedom. With increasing sample size or increasing degrees of freedom, the t-distribution converges to the *standard normal distribution*.

In Fig. 5.1, both the so-called density functions of two t-distributions and the density function of the standard normal distribution are given (dotted line: t-distribution with $df = 3$; dashed line: t-distribution with $df = 13$; solid line: standard

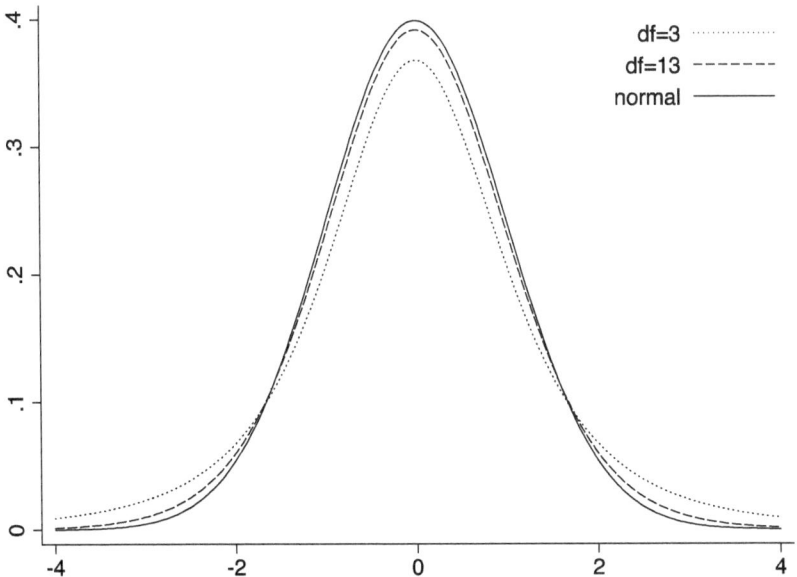

Fig. 5.1 t-distributions depending on the degrees of freedom and normal distribution. (Own representation)

normal distribution). The graphs of the density functions indicate with the area between the respective curve and a certain interval on the x-axis (e.g. $0-2$) how likely it is that a randomly drawn value from the distribution lies in this interval. For example, for the t-distributions (and also for the standard normal distribution), the area in the middle around the value 0 is relatively large, which tells us that a randomly drawn value from the distribution is relatively likely to be close to the value 0, extreme areas are unlikely.

As a reminder: The expression shown above only follows the condition of H_0 a t-distribution as shown in Fig. 5.1 or the standard normal distribution. If we now draw a random sample from the population for our study, to determine the product-moment correlation coefficient, we can imagine that we also draw a corresponding t-value from the t-distribution. If the null hypothesis actually holds, the probability that our t-value is close to the value zero is relatively large and a small t-value in absolute terms would be an indication for the H_0. However, we become suspicious of the H_0 if we get a t-value that is (in the unidirectional case) clearly above or clearly below the value zero. If the t-value is in an *extreme or critical region* of the t-distribution or standard normal distribution that applies under the null hypothesis, we decide against the null hypothesis, as it seems extremely unlikely to us to draw such an extreme value if there is no relationship between the variables age and traditional gender role orientation in the population. This also becomes clear with a look at the formula for t. If there is no relationship, r and consequently also t should be close to the value zero. But if the t-value calculated on the basis of our sample data is (in absolute terms) large and lies in a critical region of the t-distribution, we decide in reverse for the H_1. The question now is only where we draw the boundary for this critical region.

Critical region In the fifth step, we determine the critical region. We define this based on our chosen significance level. In step 3, we decided on $\alpha = 0.01$. This implies how the boundaries for the critical region of the t-distribution are to be drawn. In the unidirectional case and at a significance level of one percent, the area in the left and right extreme regions of the t-distribution must also add up to only 1%. Thus, there remains an area of 0.5% (or 0.005) on the left and right. To determine the values that delimit these areas, certain tables are available. Since we have a relatively large sample, we can assume that our test statistic follows a standard normal distribution under the assumption of the H_0 (since, as already mentioned, the t-distribution converges to the standard normal distribution with increasing degrees of freedom). For the standard normal distribution, the so-called z-table is available, which indicates which values of our test statistic delimit which area of the density function. For each z-value (or t-value), the area

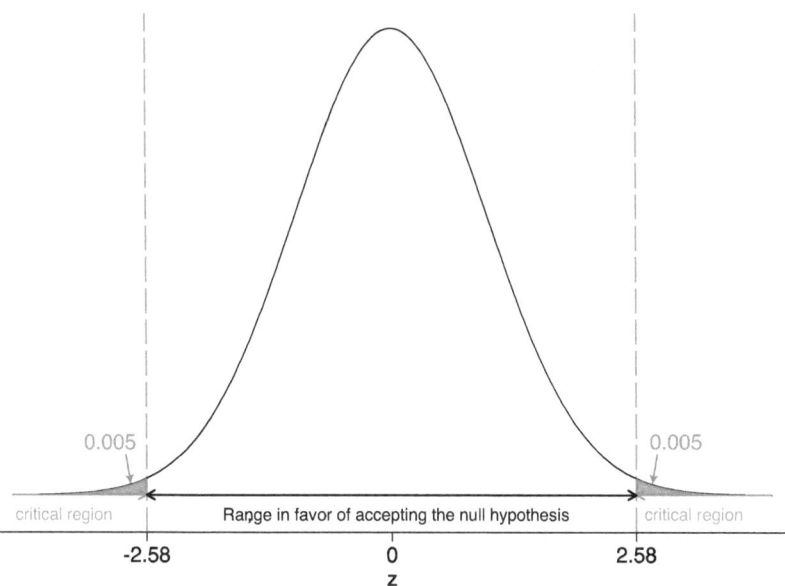

Fig. 5.2 Density function of the standard normal distribution with critical regions for two-sided testing with $\alpha = 0.01$. (Own representation)

that lies in the interval from minus infinity to the respective z-value (or t-value) under the curve of the standard normal distribution is given. The table can be found in numerous textbooks (e.g. Bortz and Schuster 2010) or is also available online, for example, from the University of Cologne (see http://eswf.uni-koeln. de/glossar/zvert.html). For our case, we need the z-values that delimit the areas of 0.005 and 0.995, so that an area of 0.005 remains on the left and right of the values. These are the values -2.58 and $+2.58$ (see Fig. 5.2).[2]

As explained in Sect. 5.2, we can also formulate directed hypotheses if we can justify them theoretically. Then, in the case of a suspected negative correlation

---

[2]The exact value 0.005 (or 0.995) for the area under the standard normal distribution is usually not listed in a z-table, instead the values 0.0049 and 0.0051 (or 0.9949 and 0.9951) and the corresponding z-values 2.58 and 2.57 (or -2.57 and -2.58) are given. We choose here the value 2.58 (or -2.58), so that the critical region is smaller and we thus hold on to the H_0 a little longer.

(H_1:$\rho<0$), we are only interested in the left area of the distribution, or in the case of a suspected positive correlation (H_1: $\rho>0$), we are only interested in the right area of the distribution (see Fig. 5.3).

In the directed case, the critical region of the standard normal distribution for a postulated negative correlation is bounded by the value -2.33 (or -2.326), for a postulated positive correlation by the value $+2.33$ (or $+2.326$). Empirically determined values for our test statistic that are found in the critical regions are each evaluated as evidence for the alternative hypothesis.

Calculation of the test statistic based on the sample data Now we calculate the test statistic using our sample data. For $r_{x,y}$ we obtain (with the help of the statistics program Stata) a value of 0.1933 (to keep the rounding error small, we give the value to four decimal places). We can insert this value and the sample size into the formula given above and thus determine t_{emp}:

$$t_{emp} = \frac{r_{x,y} \cdot \sqrt{n-2}}{\sqrt{1 - r_{x,y}^2}} = \frac{0,1933 \cdot \sqrt{1742 - 2}}{\sqrt{1 - 0.1933^2}} = 8.22$$

It should be noted again that this expression follows a t-distribution (or standard normal distribution) with the expected value 0—provided that the H_0 holds.

Decision for or against the null hypothesis Finally, we make a decision regarding the H_0. If the empirically calculated value of the test statistic lies in the critical region (see Fig. 5.2), we reject the H_0 in favor of the H_1, otherwise we retain the H_0. In our case, $\mid t_{emp} \mid > t_{krit}$, so we decide for the H_1 and assume that the found relationship is non-random and that the variables age and traditional gender role orientation vary in dependence on each other in the population.

To show that the principle of significance testing can be applied to different questions, we will discuss the procedure for testing a mean difference in the following. We will also address the sampling distribution and the standard error, two important concepts that are also essential for understanding the statistical power (see Sect. 5.8) later.

5.5   Step by Step using the Example of a Difference Hypothesis

As already indicated, the logic of the significance test and also the seven-point scheme can be applied to a whole range of different questions. For example, if we are interested in the extent to which two independently drawn groups of

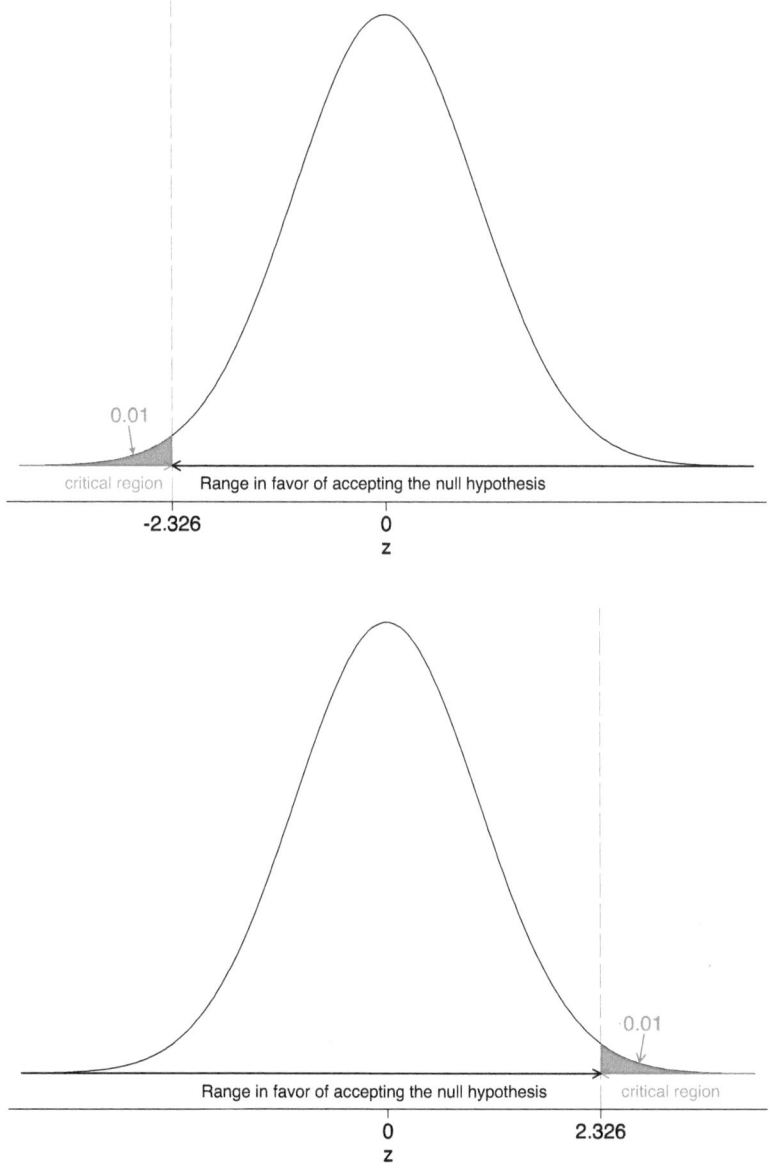

Fig. 5.3 Density functions of the standard normal distribution with critical regions for one-sided testing with $\alpha = 0.01$. (Own illustration)

persons differ with respect to a metric variable, we can perform another inference test according to the same principle as in Sect. 5.4. We again use the data from the GGSS of the year 2016 and want to investigate whether there is a difference between persons who were socialized in East Germany and persons who were socialized in West Germany with regard to their traditional gender role orientation.[3]

Assumptions In the context of our question, we examine whether two groups differ with respect to a metric characteristic. Since we can assume that the persons of the two groups (persons who were socialized in East Germany and persons who were socialized in West Germany) are not connected to each other (e.g. by kinship or similar) and the persons of these two groups were drawn independently of each other for our sample, the *two-sample t-test* is initially suitable as an inferential statistical method for this case. This parametric method is based on the assumption that the metric characteristic (in our example the traditional gender role orientation) is normally distributed in both populations. For simplicity, we assume that this assumption is met. Furthermore, the two-sample *t*-test requires that the metric characteristic has a comparable variance in both groups, i.e. that *homogeneity of variance* is present. If this assumption is not met, a modification in the form of the so-called *Welch test* should be performed. In principle, the test for homogeneity of variance can also be done according to the seven-point scheme. We skip that here, but report that the variance of the traditional gender role orientation in the two groups differs significantly.[4] Since *heterogeneity of variance* is present, we apply the Welch test to examine whether persons from West and East differ in their traditional gender role orientation.

---

[3] We speak in this context of difference and association hypotheses. However, this distinction is rather artificial, as a difference hypothesis can usually also be transformed into an association hypothesis. If we examine the difference between East Germans and West Germans with regard to their traditional gender role orientation, we are essentially examining the association between the East/West variable and the traditional gender role orientation.

[4] In SPSS, when performing a *t*-test for independent samples, the results of the Welch test are also given in the output by default. Based on the also automatically output Levene test, which checks the homogeneity of variance, it can then be decided which output to interpret. In Stata, a test for homogeneity of variance can be performed beforehand with the command robvar and then the corresponding difference test for equal or unequal variances or the Welch test with the command ttest.

Null and alternative hypothesis Based on theoretical considerations and previous empirical findings (e.g. Lois 2020; Mays 2012), we assume that people who were socialized in West Germany have a stronger traditional gender role orientation than people whose socialization took place in East Germany. Thus, we formulate a directional hypothesis. To set up our statistical hypotheses, we again use Greek letters. This makes it clear that we are referring to the population and the parameters we are interested in are population means. We denote these by μ_1 (population mean of people who were socialized in West Germany) and μ_2 (population mean of people who were socialized in East Germany). Again, we first formulate the null hypothesis.

$$H_0 : \mu_1 \leq \mu_2$$

Next, we set up the alternative hypothesis, which asserts the opposite of H_0.

$$H_1 : \mu_1 > \mu_2$$

Significance level The significance level, i.e. the probability of the alpha error, we again set at $\alpha = 0.01$. This means that we have a 1% chance of being wrong if we decide for H_1 according to our sample data.

Test statistic In this step, it is again about the question of how we can determine a test statistic whose distribution is known to us in the case of the null hypothesis. We would like to address here, as announced, also the terms of the *sampling distribution* and the *standard error*. They are central for the understanding of inferential statistics in general and also of the following subchapters.

For this, we would like to first look at the question more precisely. We are interested in whether two independent groups (East/West) differ from each other with respect to a metric characteristic (attitude towards gender roles). Of course, our interest again refers to the population, the samples are again 'only' a means to an end. Since we want to know whether the two groups differ with respect to a metric characteristic, it makes sense to use the respective group mean for the comparison. Thus, we examine whether the two groups differ from each other on average. Based on these considerations, we can now formulate our question as follows: We are interested in whether two population means differ from each other or whether two sample means differ significantly from each other. For the construction of our test statistic, which aims to compare two means, it is obvious to use the difference of the two sample means, which serves as an estimate for the difference of the population means. The sample statistic thus formed *sam-*

ple mean difference has, like other sample statistics (e.g. means, correlations), a certain distribution. This distribution is called *sampling distribution* and one can imagine this distribution in the context of a thought experiment as follows: Let us imagine that we draw the East/West samples not only once and determine the sample mean difference of the traditional gender role attitudes, but that we repeat this infinitely often. In this case, we would finally have not only one sample mean difference, but infinitely many sample mean differences, which are distributed in a certain way, i.e. there are smaller and larger differences, some value ranges occur more frequently than others. According to the *Central Limit Theorem,* an important principle of statistics, sample means and consequently also sample mean differences have a normal distribution from a certain sample size[5] (Kühnel and Krebs 2001). This normal distribution is defined like any other normal distribution by a mean and a variance. The mean of the sampling distribution, i.e. the value that one can expect on average from the sampling distribution, corresponds to the difference of the population means, i.e. the sought true value. The variance of the mentioned normal distribution is also called *standard error of the difference.* Generally speaking, the standard error is the standard deviation of a sampling distribution, i.e. the distribution of a sample statistic (e.g. sample mean difference, sample mean, sample correlation). The standard error thus indicates how strongly sample statistics vary around the true value in the population. The standard error of the difference is an indicator of how strongly all possible sample mean differences vary around the true population mean difference. If we now draw two samples from the East/West populations in reality, we can imagine that we also automatically draw a sample mean difference from the sampling distribution, i.e. a certain value from the corresponding normal distribution. Assuming that the population means do not differ, the mean of the sampling distribution is logically zero. Deviations from the mean would then only be due to the random process of sampling. The standard error of the difference is in the case of *known population variances* now defined as follows:

$$\sigma_{\bar{x}_1 - \bar{x}_2} = \sqrt{\frac{\sigma_1^2}{n_1} + \frac{\sigma_2^2}{n_2}}$$

Here, the standard deviations (σ_1, σ_2) or variances in the respective population are divided by the respective sample size (n_1, n_2). Looking at the formula, one can realize the following: The standard error of the mean difference is smaller, the lower

---

[5] Often the rule $n \geq 30$ or $n_1 \geq 30$ and $n_2 \geq 30$ is used here (e.g. Kühnel and Krebs 2001).

the variation of the characteristic in the populations and the larger the samples are. With a small standard error, the sample mean differences scatter narrowly around the true value of the mean difference in the population. If we draw two samples from the two populations under the assumption of a small standard error and calculate the sample mean difference, the probability is also high that this is close to the true value (even if we do not know the mean difference in the population at all). For the construction of our test statistic, we would now use our sample mean difference as a matter of course. If this difference is not equal to zero, this could be seen as an indication that there is a mean difference in the population—after all, the sample means serve as estimates for the population means. However, two sample means will rarely be identical, so we can only decide by means of the inference test whether we consider the difference as a random fluctuation within the scope of the sampling or as more than random and based on the populations. Furthermore, we need the above-mentioned standard error of the mean difference for our test statistic. This formula requires the knowledge of the variances in the two populations. However, since this is usually not given, we have to use the variances in our samples as an estimate for the population variances. Finally, this results in a test statistic that follows a t-distribution and is defined as follows[6] (Fig. 5.1):

$$ t = \frac{\bar{x}_1 - \bar{x}_2}{\sqrt{\frac{\hat{\sigma}_1^2}{n_1} + \frac{\hat{\sigma}_2^2}{n_2}}} $$

In the denominator, there is now the *estimated standard error,* which is based on the *estimated population variances,* which can be calculated using sample data as follows:

$$ \hat{\sigma}_1^2 = \frac{\sum_{i=1}^{n} (x_{1i} - \bar{x}_1)^2}{n - 1} \quad \text{und} \quad \hat{\sigma}_2^2 = \frac{\sum_{i=1}^{n} (x_{2i} - \bar{x}_2)^2}{n - 1} $$

Those who want to understand the (relatively simple) derivation of the test statistic t more precisely, may refer to more comprehensive textbooks (e.g. Bortz and Schuster 2010, p. 120 ff.)[7] . For us, it is sufficient at this point that we can imagine what is meant by the sampling distribution and the standard error; as well as

---

[6] In the case of unequal variances, the determination of the degrees of freedom is somewhat more complex and can be looked up, for example, in Ludwig-Mayerhofer et al. (2014) or Bortz and Schuster (2010).

[7] Bortz and Schuster (2010) use the notation s^2 for the estimation of a population variance.

that our test statistic for testing a mean difference follows a t-distribution and can be determined by dividing the sample mean difference by the estimated standard error.

Critical region We have already discussed the critical region in the context of the correlation hypothesis. However, we have now formulated a directional hypothesis, so that we are only interested in one side of the sampling distribution (see Fig. 5.3). Since we are again using the GGSS data and thus have a large sample, we can again take advantage of the fact that the t-distribution converges to the standard normal distribution as the sample size increases and read the limit of the critical region from the z-table (see http://eswf.uni-koeln.de/glossar/zvert. html). Because we assume with the H_1 that the mean of the traditional gender role orientation for persons who were socialized in West Germany is larger, we reject the null hypothesis only if t takes positive values. For a significance level of 1%, we are looking for the z-value that delimits the upper 1% of the standard normal distribution. This is the value $z = 2.33$ (or $z = 2.326$).

Calculation of the test statistic based on the sample data Now we want to calculate the empirical value for t and insert it into the formulas given above. For the values to be inserted, we have again used the GGSS data set and the statistical program Stata (to keep the rounding error low, we give the values to be inserted to four decimal places):

$$t_{emp} = \frac{1.8384 - 1.7047}{\sqrt{\frac{0.4806}{1021} + \frac{0.3936}{583}}} = 3.95$$

Decision for or against the null hypothesis Finally, we compare the empirical value of t (=3.95) with the value that delimits the critical region of the standard normal distribution (=2.33). Since here $|t_{emp}| > t_{krit}$, we reject the H_0 in favor of the H_1. We accordingly assume that persons who were socialized in West Germany have a stronger traditional gender role orientation than persons who spent most of their childhood and youth in East Germany.

5.6   Which Test Fits my Project?

There are a number of different tests available to test hypotheses. A good and comprehensive overview is provided by Clauß et al. (2017) and Hedderich and Sachs (2020). The selection of the test that suits your own research project depends, among other things, on your own research question (e.g., do two popula-tions differ with respect to a metric characteristic?), the type of sampling (e.g., independently drawn samples), the level of measurement of the variables of inter-est (e.g., nominal and metric), the number of samples, and the question of whether certain distribution assumptions (e.g., normal distribution) can be considered as fulfilled (see Clauß et al. 2017, p. 181 ff.; Hartmann and Lois 2015, p. 48 ff.).

In addition to *goodness-of-fit tests,* which compare a distribution based on sam-ple data with a theoretical distribution assumed in the population, *difference tests* comprise a relatively large group of inferential statistical methods (Clauß et al. 2017). Within the framework of testing a difference, three types of questions can be roughly distinguished. For example, we may be interested in the question of how two or more groups differ with respect to a measure of central tendency (e.g. mean) and perform a *location test.* Furthermore, we could examine whether the groups differ with respect to a measure of dispersion (e.g. variance) and apply a *dispersion test.* A so-called *omnibus test* is used when we want to examine several distribu-tion properties. These three categories can be further subdivided and we would like to illustrate this exemplarily and partially for the group of those location tests that compare *two groups* with respect to the mean (see Fig. 5.4): If we are interested in whether two populations differ with respect to a metric characteristic, independ-ent or dependent (paired) samples may be present. In the case of paired samples, the drawing of one sample is dependent on the other. For example, if 150 men are compared with their fathers with respect to the traditional gender role orientation, the fathers were drawn in dependence on their sons (or vice versa). Independent samples are spoken of, on the other hand, when the drawing of one sample is inde-pendent of the drawing of the other (see Sect. 5.5). Within the framework of the difference tests for two *independent* samples, the *two-sample t-test* can now be distinguished from the *Mann–Whitney U-test.* The latter belongs to the non-par-ametric, i.e. distribution-free methods and can be used when the prerequisites for the parametric method of the two-sample *t*-test are not met (e.g. normal distribu-tion of the metric characteristic). If there are unequal variances in the populations to be compared in the *t*-test for independent samples—which should be checked in advance by a corresponding dispersion test—the adjustment according to *Welch* is usually made, as done above. A non-parametric alternative to the parametric *paired t-test* (for dependent samples) is the *Wilcoxon test.*

If more than two groups are to be compared, the methods of *variance analysis* can be used, which can be subdivided again and in a similar way. As an alterna-tive multivariate analysis method, the *regression analysis* is finally available, into which the methods of the *t*-test and the variance analysis can be integrated rela-tively easily (see chap. 6).

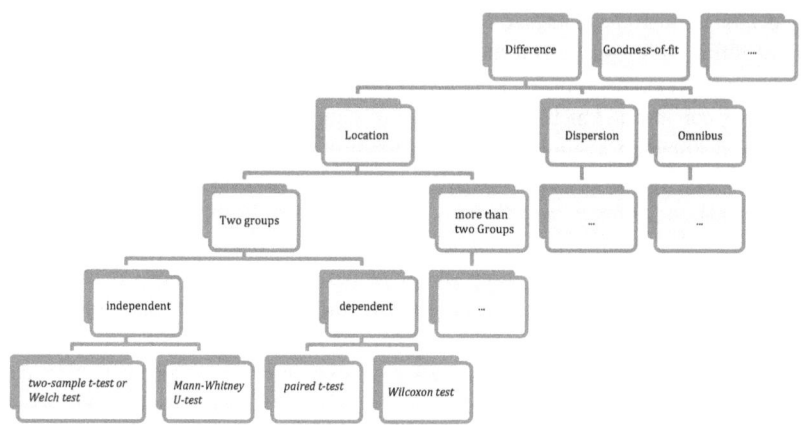

Fig. 5.4 Overview of inferential statistical methods for testing differences[8]. (Own representation)

5.7   Effect Size

It is not enough to pay attention only to the significance of a correlation or a difference when analyzing social science data. As we discussed in the section on testing the difference hypothesis, the standard error is smaller the larger the sample is. This can have the consequence that even small and (from the perspective of a social scientist) irrelevant correlations or differences are significant in large samples. That is, we can 'detect' them with the help of our inference test, but they do not have significant consequences for our or the reality of the population. For example, if we would find out on the basis of our data that people from East and West Germany differ significantly in their monthly salary, but the difference would only be 1.50 €, one can at least question whether this difference is relevant. Therefore, we should always also calculate *effect sizes*, which indicate how strong and thus how meaningful the correlation or the difference is. This also applies in reverse. That is, if the data of a relatively small sample are available, we may miss relevant correlations if we pay attention only to the significance.

---

[8] A comprehensive overview is also provided, for example, by the homepage of the methods consulting service of the University of Zurich: https://www.methodenberatung.uzh.ch/de/datenanalyse_spss.html

Depending on the test and test statistic, there are a whole range of different effect sizes available. We cannot provide an overview here, but we would like to at least determine the effect sizes for our two examples.

This is very easy for the correlation between age and traditional gender role orientation. The product-moment correlation coefficient r is already a measure of the effect size. To assess the strength of the correlation, the limits suggested by Cohen (1988) are often used. According to this, a correlation from $r = 0.10$ is to be regarded as a *weak,* a correlation from $r = 0.30$ as a *medium* and a correlation from $r = 0.50$ as a *strong* correlation. In our case, according to Cohen (1988), there is a weak correlation with $r = 0.19$. These limits are, however, only to be understood as guidelines and it is always also context-dependent whether an effect is considered weak, medium or strong.

For our difference hypothesis, various effect sizes are possible. For example, we could calculate *Cohen's d* or *Hedges g*. However, we use here the possibility to convert the t-value into r (Tarnai 1987), so that we can stick to the already mentioned guidelines of Cohen (1988):

$$r = \sqrt{\frac{t^2}{t^2 + df}}$$

If we insert into this formula, we get for our difference a value of $r = 0.11$, which indicates a weak effect.[9] The advantage of calculating r is also that by squaring the correlation coefficient, the *coefficient of determination* R^2 can be calculated (Tarnai 1987). It gives (multiplied by 100) the proportion of variance that the two variables of interest share. In the case of the correlation hypothesis, the variables age and traditional gender role attitudes share about 3.6% of their variance, in the case of the difference hypothesis, the variables East/West socialization and traditional gender role orientation have only about 1.2% of common variance.

5.8 Statistical Power

In Sect. 5.3 we explained that it is usually advisable to keep the probability of an alpha error low, so that the probability that we are wrong is small when we decide for H_1. But what if we decide against H_1 and retain H_0—how likely are we to be

---

[9] The degrees of freedom can be read from Fig. 5.6.

wrong then? In other words: How large is the probability of a beta error (type II error)? The beta error is also important because the so-called *power* can be calculated based on its probability of occurrence:

$$\text{Power} = 1 - \beta$$

The Greek letter β represents the probability of committing a beta error. If H_1 applies in the population, we can decide correctly, i.e. for H_1, or wrongly, i.e. for H_0, based on the sample data. Each decision can be assigned a probability and together these probabilities add up to the value 1—after all, only these two possibilities are available, assuming that H_1 is true. If we subtract the probability of the beta error from 1, the remainder corresponds to the probability that we decide for H_1 when H_1 applies in the population. This probability corresponds to the power of a test. If the power is high, the probability is high that we will detect a relationship (or difference) that actually exists in the population with our inference test. In Fig. 5.5 four different graphs are shown for a fictitious mean difference, each with a fictitious standard error. Shown is thus how (theoretically) infinitely many mean differences are distributed, once under the assumption of H_0 and once under the assumption of H_1. The H_1 was formulated in a directional way, so that a positive mean difference in the population is assumed. Important for the understanding of the power is that we assume for each graph that the alternative hypothesis is true and consider what happens if the test statistic falls into a certain range. The solid line thus represents the "true" distribution. The dark gray area corresponds to the significance level. If the value of our test statistic falls into this area, we decide for H_1 according to the seven-point scheme. If our test statistic is outside the critical region, we erroneously retain H_0. The light gray area corresponds to the probability of committing a beta error. If we subtract the light gray area from the area under the sample distribution of H_1, we get the power. As can be seen in the figure, the probability of the beta error in graph a) is relatively large and thus the power is relatively small. In graph b) the light gray area is much smaller, because here the standard error is smaller, i.e. the mean difference in the population is the same, but the mean differences vary less. Compared to graph a), the power in graph c) is also larger. Here, the standard error is the same, but the mean difference is larger. The mean differences are no longer distributed around the value 2, but around the value 3, so that the distribution of H_1 is "shifted" to the right. In the last graph d) both a larger effect and a smaller standard error than in graph a) are present, so that here a particularly high power can be observed. From this it should become clear that the power depends on the *effect size,* in our case on the size of the mean difference, and

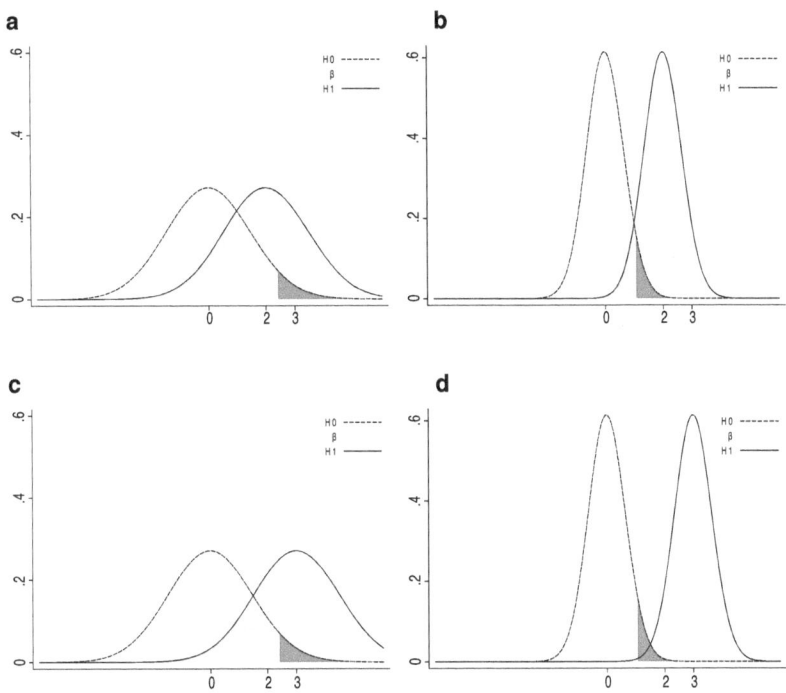

Fig. 5.5 Sampling distributions of the null hypothesis, probability of the beta error and sampling distribution of the alternative hypothesis depending on the standard error and the size of the mean difference. (Own illustration based on Bortz and Schuster 2010, p. 109; Bühner and Ziegler 2017, p. 216; Ludwig-Mayerhofer et al. 2014, p. 169)

on the *standard error,* which in turn depends on the *variation* of the characteristic in the population and on the *sample size* (see Sect. 5.5).[10]

This is (hopefully) still relatively easy to understand. However, the following problem arises: How can the probability of a beta error be determined at all?

---

[10]For the purpose of illustrating the test power, we assumed here that we know the variances of the characteristic in the groups and thus the standard error is known, so that we can represent the distribution of the mean differences as normal distributions. If this is not the case, the variances and the standard error have to be estimated; the resulting *t*-values then follow a *t*-distribution accordingly.

The H_1 is not precisely defined, unlike the H_0. In the case of the H_0, we are in the world where no correlations or differences exist, where, for example, a mean difference between two independent groups equals the value zero and where the mean differences in the sampling distribution vary around the value zero. We would need a similarly exact specification for the H_1. However, we usually do not have such a specification, which means that we also do not know by what value the mean differences are distributed in the case of the H_1. One way to deal with this dilemma is to think about how large an effect (in our case how large a difference) should be at least, so that it is worth being discovered (cf. Ludwig-Mayerhofer et al. 2014). Under this assumption, we could also estimate before the data collection how large the sample size must be, so that the test power reaches a certain level (e.g. is above 0.80). Here, a useful program is G*Power (Faul et al. 2007, 2009), which is available for free for different operating systems.

5.9   The P-Value

Statistical programs do not follow the seven-point scheme presented here in the presentation of the results of a hypothesis test, even though the applied principle is the same. Usually, only the value of the test statistic is given, but not the respective table of the corresponding distribution or the threshold values for the critical region of the test statistic. However, we do not have to compare the empirically calculated value of the test statistic with the respective critical region here, as the so-called *p-value* is output by the programs. In Fig. 5.6 the Stata output for the test of the mean difference performed in Sect. 5.5 is shown.

As can be seen in Fig. 5.6, Stata presents several p-values for different alternative hypotheses at the bottom of the output: $Pr(T<t), Pr(|T| > |t|)$ and $Pr(T> t)$. Since we formulated a directional alternative hypothesis, in which we assume that there is a positive mean difference, we are interested in the p-value on the right side. This is smaller than 0.0000 and indicates how likely it is that we obtain the output value of the test statistic or an even more extreme value under the condition of the null hypothesis. If the p-value is *smaller* than the significance level we chose, there is a significant result, as in this case it is clear that the test statistic lies in the critical region. Accordingly, we decide for the H_1. If the p-value is greater than or equal to the significance level, we retain the H_0. This principle also applies to other inference tests and their outputs.

```
Two-sample t test with unequal variances
```

Group	Obs	Mean	Std. Err.	Std. Dev.	[99% Conf.	Interval]
0	1,021	1.838394	.0216962	.6932616	1.782403	1.894384
1	583	1.704688	.0259832	.6273736	1.63754	1.771837
combined	1,604	1.789796	.0168027	.6729475	1.746464	1.833129
diff		.1337053	.0338504		.0463857	.221025

```
    diff = mean(0) - mean(1)                                      t =    3.9499
Ho: diff = 0                          Welch's degrees of freedom =    1314.55

   Ha: diff < 0                   Ha: diff != 0                   Ha: diff > 0
Pr(T < t) = 1.0000         Pr(|T| > |t|) = 0.0001         Pr(T > t) = 0.0000
```

Fig. 5.6 Stata output of the *t*-test for testing the mean difference between East (Group 1) and West (Group 0) with respect to the traditional gender role orientation. (Source: GGSS Cumulation 1980–2018, survey year 2016)

5.10   The Relationship of Hypothesis Testing to Confidence Intervals

At the beginning of Sect. 5.2 we explained that there are roughly two areas in inferential statistics that allow us to learn something about the population: estimation and testing. We have discussed *testing* hypotheses in detail. In *estimation*, point and interval estimators can be distinguished. For example, if we want to estimate the difference between two population means and calculate the difference between two sample means, we have performed a *point estimation*.[11] If we instead calculate an interval that contains the difference of the population means with a certain probability, for example with a probability of 95% or 99%, we perform an *interval estimation*. In Fig. 5.6 (*t*-test output) the 99% *confidence interval* for the mean difference of the people from East and West Germany regarding the traditional gender role orientation can be seen. This confidence interval indicates that the interval with the limits (to two decimal places) 0.05 and 0.22 contains the

---

[11]We have already done this in the context of calculating the test statistic *t* in Sect. 5.5.

mean difference in the population with a probability of 99%.[12] Here too, we do not know what reality actually looks like and whether the estimated interval really contains the true mean difference (for the logic and calculation of confidence intervals see Ludwig-Mayerhofer et al. 2014 or Kühnel and Krebs 2001).

The confidence interval shows how precisely the parameter can be estimated. The results of interval estimation and hypothesis testing are not independent of each other, so that it can be inferred from the confidence intervals whether a certain hypothesis test is significant or not. As a rule, for an *unidirectional* hypothesis, the following applies: If the confidence interval contains the value postulated under the null hypothesis for the population parameter (in our case this would be the value 0), the null hypothesis is retained, otherwise it is rejected in favor of the H_1. For *directional* hypotheses, it must be taken into account that the critical region lies only on one side of the sampling distribution (the confidence interval to be used for the decision is then smaller) and that the value of the test statistic must have the sign that matches the H_1 in order to speak of a significant result (for a more detailed presentation regarding confidence intervals and directional hypotheses and also regarding the peculiarity of confidence intervals for proportion values see Ludwig-Mayerhofer et al. 2014).

5.11   Afterword

As mentioned at the beginning of this chapter, it is not easy to understand the principle of the significance test. Possibly, it may also be enough at first to remember the seven-point scheme and to keep in mind when interpreting the outputs of statistical programs that a significant result is present when the displayed p-value is smaller than the pre-selected significance level. So anyone who has not completely understood the hypothesis testing after reading the chapter once should not be discouraged, possibly read the chapter again and also consult further literature. We have presented the concept of the significance test and central terms such as Central Limit Theorem, sampling distribution and standard error very compactly. The advantage of such a compressed introduction is that many concepts on which the hypothesis test is based are explained within one chapter

---

[12] The confidence interval does *not* indicate that the desired parameter lies exactly within these limits with a probability of 99%, but only that 99% of all possible confidence intervals cover the desired parameter and 1% do not.

and not between different chapters on topics such as probability theory, random variables and their distribution, estimation and testing. However, it is certainly beneficial for a deeper understanding to also review the chapters of more extensive textbooks on this topic after such a short introduction. We recommend, for example, the also relatively compact presentation by Ludwig-Mayerhofer et al. (2014), the corresponding chapters of the work by Bortz and Schuster (2010) or the also relatively detailed explanations by Kühnel and Krebs (2001). A impressive overview of a variety of different inference tests is provided by Hedderich and Sachs (2020). As English-language literature we recommend Mohr (1990) as well as Witteloostuijn and Hugten (2022). Without going into detail, it should finally be mentioned that with Bayesian statistics there is also an alternative to the procedure described here (see e.g. Tschirk 2019).

References

Bortz, Jürgen., und Christof Schuster. 2010. *Statistik für Human- und Sozialwissenschaftler.* Berlin: Springer. https://doi.org/10.1007/978-3-642-12770-0.

Bühner, Markus, und Matthias Ziegler. 2017. *Statistik für Psychologen und Sozialwissenschaftler. Grundlagen und Umsetzung mit SPSS und R.* Hallbergmoos: Pearson.

Clauß, Günter., Falk-Rüdiger. Finze, und Lothar Partzsch. 2017. *Grundlagen der Statistik für Soziologen, Pädagogen, Psychologen und Mediziner.* Haan-Gruiten: Verlag Europa-Lehrmittel Nourney, Vollmer GmbH & Co. KG.

Cohen, Jacob. 1988. *Statistical power analysis for the behavioral sciences.* Hillsdale: Lawrence Erlbaum Associates.

Faul, Franz, Edgar Erdfelder, Axel Buchner, und Albert-Georg Lang. 2009. Statistical power analyses using G*Power 3.1: Tests for correlation and regression analyses. *Behavior Research Methods* 41:1149–1160. https://doi.org/10.3758/BRM.41.4.1149.

Faul, Franz, Edgar Erdfelder, Albert-Georg. Lang, und Axel Buchner. 2007. G*Power 3: A flexible statistical power analysis program for the social, behavioral, and biomedical sciences. *Behavior Research Methods* 39:175–191. https://doi.org/10.3758/BF03193146.

Hartmann, Florian G., und Daniel Lois. 2015. *Hypothesen Testen. Eine Einführung für Bachelorstudierende sozialwissenschaftlicher Fächer.* Wiesbaden: Springer Gabler. https://doi.org/10.1007/978-3-658-10461-0.

Hedderich, Jürgen., und Lothar Sachs. 2020. *Angewandte Statistik: Methodensammlung mit R.* Berlin: Springer. https://doi.org/10.1007/978-3-662-62294-0.

Krampen, Günter. 1979. Eine Skala zur Messung der normativen Geschlechtsrollen-Orientierung (GRO-Skala). *Zeitschrift für Soziologie* 8:254–266. https://doi.org/10.1515/zfsoz-1979-0304.

Kühnel, Steffen-M., und Dagmar Krebs. 2001. *Statistik für die Sozialwissenschaften: Grundlagen, Methoden, Anwendungen.* Reinbek bei Hamburg: Rowohlt Taschenbuch. https://doi.org/10.1007/s11577-003-0033-5.

Lois, Daniel. 2020. Gender role attitudes in Germany, 1982–2016: An age-period-cohort (APC) analysis. *Comparative Population Studies* 45:35–64. https://doi.org/10.12765/CPoS-2020-02.

Ludwig-Mayerhofer, Wolfgang, Uta Liebeskind, und Ferdinand Geißler. 2014. *Statistik. Eine Einführung für Sozialwissenschaftler*. Weinheim: Beltz.

Mays, Anja. 2012. Determinanten traditionell-sexistischer Einstellungen in Deutschland – Eine Analyse mit Allbus-Daten. *Kölner Zeitschrift für Soziologie und Sozialpsychologie* 64:277–302. https://doi.org/10.1007/s11577-012-0165-6.

Mohr, Lawrence B. 1990. *Understanding Significance Testing*. Sage University Papers: Quantitative Applications in the Social Sciences. London: Sage Publications.

Tarnai, Christian. 1987. *Einführung in die Grundlagen der Statistik*. Münster: Institut für sozialwissenschaftliche Forschung e. V.

Tschirk, Wolfgang. 2019. *Bayes-Statistik für Human- und Sozialwissenschaften*. Berlin: Springer. https://doi.org/10.1007/978-3-662-56782-1.

Witteloostuijn, Arjen van, und Joeri van Hugten. 2022. The State of the Art of Hypothesis Testing in the Social Sciences. *Social Sciences & Humanities Open* 6 (1): 100314. https://doi.org/10.1016/j.ssaho.2022.100314.

Linear Regression

6

Anyone who has read or even worked through this book up to this point and is also somewhat familiar with the empirically oriented sociological literature will have found only relatively few overlaps so far and might even be a little disappointed in view of the objective of this book. Thus, the descriptive and bivariate methods discussed in Chap. 4 serve more to get to know the data and to prepare a real social scientific data analysis than to generate insights into social processes themselves. In published texts, such (preparatory) methods, which include, for example, factor and reliability analyses, are rarely found, if at all, then usually only in brief form, as they rarely pursue an independent research interest.

At the end of this chapter, however, this assessment should have changed fundamentally, because in the following, the method of linear regression will be introduced, which is capable of testing relationships between various variables and also depicting mediated or suppressed influences (see also the explanations on the logic of data analysis in Chap. 7).[1] The main goal of a linear regression is to trace the values of a metrically measured variable back to the differences of certain other variables.

To introduce this method in an understandable way, the example of traditional gender role orientation will be used again and (explanatory) factors will be sought that influence this attitude. Even without much reflection on sociological theories,

---

[1] The ultimate goal of empirically oriented social research is always to analyze causal, i.e., causal processes. It is extremely difficult to actually depict causal processes (see, for example, Bunge 2010; Gangl 2010; Hedström and Ylikoski 2010; Opp 2010). For everyday use, one should always be aware of whether the empirically observable influencing variables result from a theoretical model (see for this distinction to a so-called variable sociology Esser 1987).

© The Author(s), under exclusive license to Springer Fachmedien Wiesbaden GmbH, part of Springer Nature 2023
F. Hartmann et al., *Social Science Data Analysis*,
https://doi.org/10.1007/978-3-658-41230-2_6

one can assume that the attitude towards what is a gender-specific appropriate behavior depends on various factors: for example, age, gender, or socio-economic status.[2] One of the great advantages of linear regressions is that they can take into account the influence of all these (and other conceivable and hopefully operationalized) variables together.

6.1   Basic Logic of Bivariate Regression

However, to get to know the method first, only one explanatory or independent variable should be considered in a first step: the age of the respondents. The theoretical assumption is the simple hypothesis that traditional gender role orientation should increase with increasing age. In Fig. 6.1 you can find a corresponding scatter plot to assess the relationship.[3]

Already based on the graph, the tendency becomes apparent that traditional gender role orientation increases with increasing age—the two variables covary positively with each other and the corresponding correlation (for the total sample) is 0.17.[4] As already demonstrated above with the different measures of association, the question now arises whether one can predict the traditional gender role orientation better if one knows the age of the persons. Bivariate or simple linear regressions are suitable for this question. However, it should be noted as a limitation that only linear functions are allowed as prediction rules. For the graph and also for the following regression models, we have centered the variable age around its mean. This means that the average age of the sample was subtracted

---

[2] In concrete research practice, these empirically suspected influencing factors must be derived from a theoretical model—and not just simply listed as here. In most cases, relatively simple and highly simplified models that capture the complex everyday situation are sufficient, such as those provided by various action-theoretical considerations that are summarized under the keyword "rational choice" (see introductory Hill 2002).

[3] To generate this figure, the data was given 'random noise' with the option 'jitter' of the Stata command "twoway scatter", i.e. random variations around the actual values. This makes the individual points easier to see. For the same reason, only a small subsample ($n = 100$) of the data set was drawn. The further analyses will be calculated with all cases, however.

[4] The total sample for the calculation of the correlation and for all following calculations consists of those cases that have a valid value on all variables listed in this chapter. This allows the different regression models to be compared well with each other, since the calculations are always based on the same cases, i.e. the sample composition does not change.

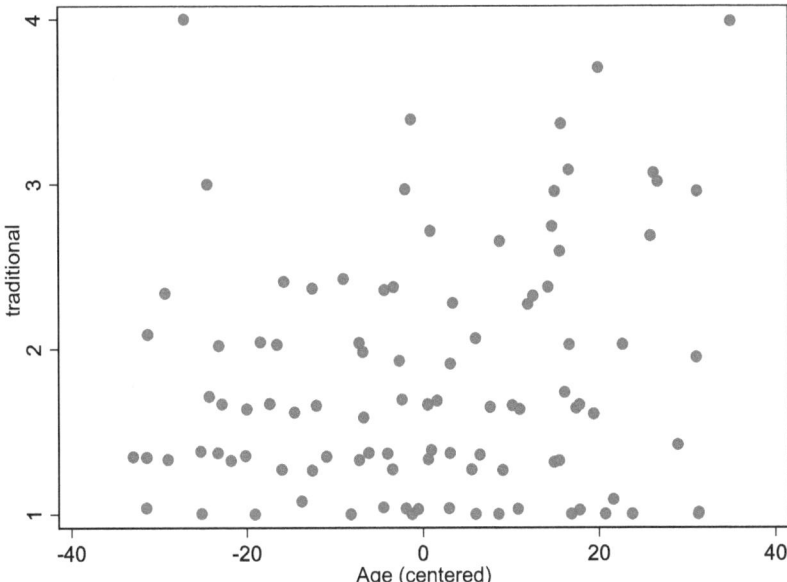

Fig. 6.1 Relationship between the (centered around the mean) age and the traditional gender role orientation. (Source: GGSS 2016)

from each age indication, so that positive values of the mean-centered variable indicate that a person is older than the average and negative values that a person is younger than the average of the sample. The value zero of the mean-centered variable corresponds to the mean of the original variable, i.e. the average age of the sample. We will later, when it comes to interpreting the constant (or intercept) of the regression model, learn why mean-centering of independent variables can be useful in the context of regression models—but nothing changes in the relations to the dependent variable, i.e. the correlation between the original age variable and the traditional gender role orientation is also 0.17.

   To understand the method well in its details, the individual analyses will be calculated in a first step with a small example data set, where the possibility exists to follow the analyses in detail with relatively little effort. In a second step, the results for the complete data set will be presented. This example data set consists of 10 persons, whose characteristics in the variables of interest can be taken from Table 6.1 and which can easily be entered into corresponding data analysis programs.

Table 6.1 Age and traditional gender role orientation. (Source: fictitious data)

Serial number	Age (centered)	Trad. orientation (scale 1–4)
1	−14	1
2	9	2
3	−3	4
4	14	3
5	21	1
6	−2	2
7	−14	1
8	−5	3
9	4	2
10	−10	1
Arithmetic means	0	2

To investigate the question of whether the knowledge of the age of the persons improves the estimation of their gender role orientation, one first needs a basic estimation that can be improved at all. For this purpose, the arithmetic mean of the gender role orientation is used.

In a second step, one has to be clear about how to define the error that one makes with this prediction. Here, a whole range of possibilities is conceivable, for various reasons the *square of the deviation* of the real gender role orientation from this mean value is considered as the error. As can be easily calculated, the mean value is 2. In Table 6.2 the error is calculated that results from predicting a traditional gender role orientation of 2 for each person.

The total sum of these squared deviations, i.e. the total error, amounts to 10. We are now looking for a prediction rule that produces a smaller error, with the restriction that only linear prediction rules are applied. So we are looking for a straight line in the simple bivariate case considered here that generates less error—in the definition just determined—than the use of the mean. If one considers that straight lines always follow the equation

$$y = b_0 + b_1 \cdot x_1$$

this is ultimately a minimization problem, which can be solved with the help of school mathematics (see for a formal presentation Urban and Mayerl 2008, pp. 46 ff.). Ultimately, one tries to minimize the squared deviations of the actual expression of gender role orientation from the predicted gender role orientation

Table 6.2 Error in predicting the traditional gender role orientation by the mean value. (Source: fictitious data)

Serial number	Prediction	Trad. orientation	Squared error (square of the deviation)
1	2	1	1
2	2	2	0
3	2	4	4
4	2	3	1
5	2	1	1
6	2	2	0
7	2	1	1
8	2	3	1
9	2	2	0
10	2	1	1

based on age. Even though in practice the calculation of the concrete values of b_0, the *intercept,* and b_1, the *slope,* probably never occurs by hand, the corresponding formulas should be given here once.

$$b_1 = \frac{\frac{1}{n} \cdot \left(\sum_{i=1}^{n} (x_i - \bar{x}) \cdot (y_i - \bar{y}) \right)}{\frac{1}{n} \cdot \left(\sum_{i=1}^{n} (x_i - \bar{x})^2 \right)}$$

$$b_0 = \bar{y} - b_1 \cdot \bar{x}$$

It can be seen that the slope of the line, b_1, is mainly determined by the covariance—with a strong positive correlation between the variables, the best prediction line is (not surprisingly) also relatively steep and positive, with a negative correlation and thus a negative covariance, the curve slope is also negative. This covariance is weighted or 'normalized' by the variance of x.

If one applies these formulas to the 10 persons from Table 6.1 that are examined more closely here, one obtains as the slope of the line b_1 the value 0.016 and as the intercept b_0 the value 2.[5] In most cases, the predictions are still not error-

---

[5]The intercept or the constant corresponds here to the average of the traditional gender role orientation, since we have previously subjected the age to a mean centering. Anyone who is interested in calculating the coefficients for the presented example data should note that the

Table 6.3 Errors in predicting the traditional gender role orientation by age. (Source: fictitious data)

Serial number	Age (centered)	Prediction	Trad. orientation	Squared error
1	−14	1.78	1	0.61
2	9	2.14	2	0.02
3	−3	1.95	4	4.20
4	14	2.22	3	0.61
5	21	2.34	1	1.80
6	−2	1.97	2	0.00
7	−14	1.78	1	0.61
8	−5	1.92	3	1.17
9	4	2.06	2	0.00
10	−10	1.84	1	0.71

free, not all points lie on the modeled line. In Table 6.3 the predicted values based on the regression line and the resulting errors are now entered for the 10 cases.

As the sum of squared errors (SSE) we get a value of 9.73. If we use the age of the respondents to predict the traditional gender role orientation, the number of error points decreases from 10 by 0.27. The prediction improves slightly. About 2.7% of the original errors can be avoided by taking age into account.[6]

This just described procedure represents the most important points of a bivariate regression, which—since the errors are defined as deviation squares—is also called OLS regression, where OLS stands for **o** rdinary **l** east squares. In the simple and here considered bivariate case, one looks for the regression line that produces the least errors, or the smallest sum of the deviation squares. The relative

---

covariance or the numerator of the fraction at b_1 in this example is 2.22 and the variance of x 140.44. The means of the two variables can be found in Table 6.1.

[6] If the data from Table 6.1 are entered into a statistics program, performing a regression results in a reduction of the error by 3.2%. These and other deviations from the values calculated here are due to rounding errors, as we always calculated with as many decimal places as indicated in the text, so that individual calculations can also be followed with a calculator. Overall, these differences can be neglected.

improvement of the error is called the *coefficient of determination* R^2. This procedure follows the logic of the proportional reduction of error (PRE).

When interpreting the regression line, one now assumes that the analyses reflect real empirical causal processes—in our case, that a higher age leads to a higher traditional gender role orientation. Not always is the direction of effect so clear as in our example, since here at least it can be assumed that the gender role orientation does not affect the age. Often, however, it cannot be ensured whether a process really works from x to y or whether the causality runs in the opposite direction, i.e. from y to x—or whether certain other variables influence both processes; such questions cannot be clarified with a simple regression alone (see also Chap. 7 or the contribution by Brüderl 2010). Despite all the technical possibilities that still need to be clarified, one must be aware of the fact that empirical data can only serve as a test for theoretical considerations, but never replace them. The first step of an empirical analysis must therefore always be the theoretical modeling and the elaboration of falsifiable hypotheses.

Similar to the logic presented in Chap. 4 for the analysis of cross tables using χ^2-tests, it is possible to test the strength of the relationship against purely random processes. This verification is done in two steps: In a first step, the overall quality of the model is considered. For this purpose, a so-called F-test is calculated, which takes into account the explained and the unexplained error sum, the number of variables used for the explanation, and the number of cases.[7] In our small example data set with 10 people, the F-value is 0.261 and is—as can be seen from corresponding tables or the outputs of the data analysis programs—not significant. Similarly, one can check whether the regression coefficient b_1 is due to random

---

[7] In the simple case of a bivariate regression that has been the focus so far, the number of explanatory factors is naturally 1—this is not always the case, as will be seen shortly. For this F-test, the so-called means of squares are first formed by dividing the errors avoided by the regression by the number of explanatory factors, or by subtracting the number of explanatory factors from the degrees of freedom of the overall analysis, which results from the number of cases minus one, and dividing the remaining errors by this. In this formation of the means of squares, the degrees of freedom of the overall model are thus divided between the explained and unexplained variation. These two quantities are divided by each other and yield an F-value. This is a very simple test that ultimately only checks whether the results are due to purely random fluctuations of the data. If the F-test indicates that the model is not significant, one should completely refrain from a substantive interpretation. This test is output in SPSS in the so-called ANOVA block (**analysis of variance**) and in Stata without any special requirement.

Table 6.4 Model fit for the relationship between age and traditional gender role orientation (Anova-block). (Source: GGSS 2016, $n = 1,458$)

	Sums of squares	df	Mean	F-value
Explained by regression	21.4128	1	21.4128	45.37
Residuals (unexplained)	687.2371	1456	0.4720	
Total	708.6499	1457		

processes.[8] In our simple model, this is also not significant. Thus, our model is not able to explain the variance of the dependent variable. However, in practice we usually deal with larger samples and the result of a significance test is, as we have already learned in Chap. 5, always also dependent on the sample size, so that both tests could turn out differently, i.e. significantly, with a larger sample. As we will see shortly, this is also the case based on the data from the GGSS.

6.2   Bivariate Regression: An Example from Practice

After the basic logic of bivariate regressions has become clear on the fictitious data set with 10 cases, the relationship between age and traditional gender role orientation will be examined using the entire GGSS data. Both in SPSS and in Stata, one obtains an analysis of the explained and unexplained variations and the F-test just discussed. For the GGSS data, the following result is found:

The F-value in Table 6.4 is significant at a significance level of 1%. With regard to the explanatory power of the model, a coefficient of determination R^2 of 0.030 is obtained.

For the factors of the regression line b_0 and b_1 the values 1.821 and 0.007 are obtained (see Table 6.5).

Of course, the corresponding indication of significance can also be found for the t-value. As a rule of thumb, one can assume that t-values greater than 2 or less than -2 are significant. Based on the F-test, one can therefore conclude whether

---

[8] For the regression weight b_1, a standard error (S.E. = standard error) is also calculated, where the correlation between the independent and dependent variable and the number of cases play a role. The quotient of the effect size b_1 by this standard error yields a test statistic t, whose significance can be checked (see Chap. 5). These tests are also output without additional requirements in SPSS and Stata.

Table 6.5 Determination of the regression line. (Source: GGSS 2016, $n = 1,458$)

	Coefficient	Standard error (S.E.)	t-value
Intercept b_0	1.821	0.018	101.18
Slope of the line b_1	0.007	0.001	6.74

the effects in this model can be attributed to purely random processes or not for the entire model and based on the t-tests for the two factors b_0 and b_1. In the present case, it can be assumed that the results are not purely random, but are due to the conditions in the population. The regression line, which reduces the error rate by 3.0% (see R^2) or explains 3.0% of the variance of the dependent variable, is:

$$y = 1.821 + 0.007 \cdot x_1$$

This also clarifies how the intercept b_0 and the slope b_1 can be interpreted. Generally speaking, b_0 corresponds to the value of the dependent variable when the independent variable is zero. In our case, the independent variable is the mean-centered age; the value zero of this variable corresponds to the mean of the original age variable. Thus, the value 1.821 is the estimated value for the traditional gender role orientation of a person who is average in age. If we had not performed mean centering, we would obtain the value 1.455 for the intercept. Formally, this would be the estimated value for the traditional gender role orientation of a person who is zero years old. Content-wise, this makes little sense, as with many variables, so we would not interpret such a value. If the value zero of an independent variable is not meaningful, mean centering is a suitable means; otherwise, mean centering is not necessary. The slope b_1 is unaffected by this. Again, generally speaking, the value of b_1 indicates by how much the dependent variable increases when the independent variable increases by one unit. In our case, this means: with each additional year, the traditional gender role orientation increases by the value 0.007.

6.3 "You'll Never Walk Alone"—Multivariate Regression

To explain an interesting sociological variable, the reference to a single independent variable is never sufficient in general—sociological processes are multicausal, different factors determine, for example, the traditional gender role orientation and the actually interesting question is how strong the individual factors work. To this end, in a further step, the influence of the father's occupational status is

Table 6.6 Determination of the regression plane. (Source: GGSS 2016, $n = 1{,}458$)

	Coefficient	S.E	t-value
Intercept b_0	1.821	0.018	102.69
b_1: Influence of age	0.006	0.001	5.64
b_2: Influence of father's ISEI	−0.006	0.001	−6.69

to be taken into account in addition to the age variable. This is represented in the GGSS with the International Socio-Economic Index of Occupational Status (ISEI), which takes into account income and education (variable fisei08).[9] We assume that the traditional gender role orientation is lower with a higher status of the father. If one now uses the socio-economic status of the father in addition to the age to explain the gender role orientation, one obtains an R^2 value of 0.059 for a significant overall model (or a corrected R^2 of 0.058; the explanation of the correction follows below). Instead of a regression line, a regression plane is now predicted, whereby the method of least squares is also used here. A corresponding plane can be captured by the following equation:

$$y = b_0 + b_1 \cdot x_1 + b_2 \cdot x_2$$

In our example, the values for the different factors b_0, b_1 and b_2 result in Table 6.6. Since the ISEI cannot take the value zero, we have also centered this variable around the mean, so that the intercept b_0 can be interpreted sensibly.

The intercept is again the estimated value of y, when the independent variables are zero. Since we have mean-centered both independent variables, the constant of 1.821 is thus the estimated value for the traditional gender role orientation of a person who is of average age ($x_1 = 0$) and whose father has an average ISEI ($x_2 = 0$). With regard to the effects of the independent variables, it should be considered that in empirical analyses the variables used often have completely differ-

---

[9] Theoretically, it would make more sense to include the ISEI of the mother as an independent variable in the model. However, this leads to a significant reduction in the sample size, which is why the ISEI of the father is used here. The regression model that is calculated at the end of this chapter was also calculated with the ISEI of the mother, with the result that a similar proportion of the variance of the gender role orientation can be explained (R^2 for ISEI of the father: 10.68%, R^2 for ISEI of the mother: 10.72%, each with $n = 829$). If the ISEI of the mother and the ISEI of the father are included in the model at the same time, the ISEI of the mother has a stronger effect.

ent scales or that the distribution of values on the scales can be very different. If one is then interested in comparing different influences in terms of the strength of their influence (e.g. age vs. ISEI), the consideration of the regression weights can be misleading. This also becomes clear from the fact that, for example, the *b*-coefficient of age differs when age is not measured in years, but in months, although the age effect itself is the same. To avoid these problems and to establish comparability, so-called *standardized* coefficients, which are also referred to as β-coefficients (pronounced beta-coefficients), can be calculated (this naming is not uniform; see Urban and Mayerl 2008, p. 71). These quantities (in the bivariate case) are determined as follows:

$$\beta_i = b_i \cdot \frac{s_{x_i}}{s_y}$$

The respective regression coefficients are thus weighted by the quotient of the standard deviation of the respective variable x_i and the dependent variable y. If one calculates these standardized regression coefficients for the model just presented, one obtains the results in Table 6.7.

The corresponding β-coefficients have the advantage that they vary between -1 and $+1$ and are comparable in size for the individual independent variables. In Table 6.7 it can be seen from the β-coefficients that the influence of the father's ISEI is higher than the influence of the age of the respondent (so the magnitude of the values is decisive for the strength of the effect and not the sign)—if we had compared the unstandardized regression weights, we would have come to the *wrong* conclusion that the effects of the independent variables are equally strong. The difference in terms of the strength of the effect is therefore due to the different standard deviations of the independent variables (see formula for calculating β). It is important to note that these standardized regression coefficients should only be used to compare different magnitudes of influence within a sample. For different samples, the variances in the samples may differ and thus produce different β-values, which are not to be interpreted as substantive differences (see more detailed Urban and Mayerl 2008, p. 79).

Table 6.7 Determination of the coefficients. (Source: GGSS 2016, $n = 1{,}458$)

	b-coefficient	β-coefficient
Effect of age	0.006	0.145
Effect of father's ISEI	−0.006	−0.172

By including additional variables, the respective coefficient of determination R^2 and thus the quality of the model almost always increases. In order to still be able to assess the quality of the model reasonably and also to obtain a better estimator for the relationships in the population (see Diaz-Bone 2006, p. 219), a corrected coefficient of determination is often also reported, which is calculated as follows:

$$R^2_{\text{adjusted}} = R^2 - \frac{(1 - R^2)}{n - j - 1}$$

Here, n denotes the number of cases and j the number of predictors b_1 to b_j in the model. In our case, the corrected and the uncorrected coefficient of determination differ only slightly: 0.058 (corrected) vs. 0.059 (uncorrected).

The great advantage of multivariate regressions is that they allow to control the influence of different independent variables simultaneously. Thus, the effect of age described above of 0.006 indicates this influence under control of the level of the ISEI variable. This means that in the context of the multivariate regression, the effect of age is determined by taking into account the effect of the ISEI—the ISEI is thus kept constant, so to speak. Of course, this also applies vice versa for determining the influence of the father's ISEI, taking into account the effect of age. How this is technically to be understood is presented in a brief excursion.

Excursion: Control of third variables The basic idea of multivariate methods is to determine the influence of a certain independent variable while controlling for other independent variables. In our example, the effect of the father's ISEI on the traditional gender role orientation is to be clarified while controlling for age. To better understand how this is done, three steps are necessary: In a first step, the influence of age on the traditional gender role orientation must be examined. As shown above (see Table 6.5), the following regression line results from the GGSS data:

$$\text{Trad. orientation} = 1.821 + 0.007 \cdot \text{age}.$$

Using this line, the prediction of the traditional gender role orientation is improved by about 3%. It becomes clear, however, that the gender role orientation does not depend solely on the age of the persons, because 97% of the variation remains unexplained. This part cannot be attributed to fluctuations in age. The traditional gender role orientation shows unexplained fluctuations, so-called *residuals*.

However, there may also be a relationship between the independent variables, i.e. between the age of the respondents and the ISEI of their fathers. In a second step, we also perform a bivariate regression here first and determine the residuals of the ISEI under control of the age. Using the GGSS data, the following equation is determined for this purpose:

$$\text{ISEI of the father} = 0 - 0.201 \cdot \text{age}.$$

This improves the prediction of the father's ISEI by 2.7%. If one now performs a bivariate regression of the residuals of the traditional gender role orientation on the residuals of the father's ISEI in a third and final step, one obtains the influence of the father's ISEI on the gender role orientation under control of the age. The following value for the slope results from this.[10]

$$\text{Trad. orientation}_{\text{controlled for age}} = 0 - 0.006 \times \text{ISEI of the father}_{\text{controlled for age}}$$

Both variables of interest are thus controlled for the influence of age and the corresponding values, the residuals, are linked with each other using a simple linear regression model. Subsequently, of course, the influence of age on the gender role orientation, which is adjusted (or controlled) for the influence of the ISEI, is also examined. Together, this results in the values reported in Table 6.7.

To be able to understand this procedure concretely, we shall refer back to the small data set with 10 persons that was already presented above. Here, the ISEI of the father is also taken into account for each person. The corresponding data set is given in Table 6.8.

As introduced above, two bivariate regressions have to be calculated first, in order to obtain the values adjusted for the influence of age with regard to the traditional gender role orientation and the ISEI of the father. For this purpose, the following regression equations are determined based on the data in Table 6.8:

$$\text{Trad. orientation} = 2 + 0.016 \times \text{age}.$$
$$\text{ISEI of the father} = 0 - 0.150 \times \text{age}.$$

Based on these equations, certain predicted values and corresponding residuals result for the individual persons. These values are entered in Table 6.9.

If one performs a bivariate regression between the residuals, one obtains the following result:

$$\text{Trad. orientation}_{\text{controlled for age}} = 0 - 0.017 \times \text{ISEI of the father}_{\text{controlled for age}}$$

---

[10] Here, too, the constant takes the value zero, since it is a regression of two residuals whose mean value is 0 by definition.

Table 6.8 Age, ISEI of the father and traditional gender role orientation. (Source: fictitious data)

Serial number	Age (centered)	ISEI of the father (centered)	Trad. orientation
1	−14	−6	1
2	9	−9	2
3	−3	1	4
4	14	−15	3
5	21	4	1
6	−2	24	2
7	−14	0	1
8	−5	−4	3
9	4	1	2
10	−10	4	1

Table 6.9 Values predicted by age and residuals. (Source: fictitious data)

ID	Trad. orientation			ISEI of the father		
	Value	Prediction	Residual	Value	Prediction	Residual
1	1	1.78	−0.78	−6	2.10	−8.10
2	2	2.14	−0.14	−9	−1.35	−7.65
3	4	1.95	2.05	1	0.45	0.55
4	3	2.22	0.78	−15	−2.10	−12.90
5	1	2.34	−1.34	4	−3.15	7.15
6	2	1.97	0.03	24	0.30	23.70
7	1	1.78	−0.78	0	2.10	−2.10
8	3	1.92	1.08	−4	0.75	−4.75
9	2	2.06	−0.06	1	−0.60	1.60
10	1	1.84	−0.84	4	1.50	2.50

To finally obtain the net effect of age, the values of gender role orientation and age that are controlled for the ISEI have to be determined and related to each other.

6.4   Dummy Variables

It is of course possible to include further metric variables in such regression analyses and thus extend the explanatory approach. Here, different strategies can be pursued, for example, to find hidden relationships or to empirically explain causal mechanisms. These strategies are discussed in more detail in Chap. 7. At this point, the analyses should be extended in another direction, because often theoretically also influencing factors can be assumed that are not metrically measurable, but represent so-called qualitative variables (Urban and Mayerl 2008, pp. 275 ff.). For example, in the case under consideration here, it is certainly conceivable that gender role orientation depends on gender or religious affiliation. A great advantage of multivariate regression models is that these variables can also be integrated into these models. For this purpose, so-called *dummy variables* are introduced into the models. Dummy variables indicate whether a certain characteristic is present or not, whereby the values 1—if this characteristic is present— or 0—if this characteristic is not present—are assigned. For variables that were already dichotomously collected, as is usually the case with gender, this is easy. Then the effect of the value 1 compared to the value 0 of this variable is examined. Whether the value "man" or "woman" is assigned the value 1 is irrelevant. For qualitative data with several values, such as religious affiliation (or also gender, if more than two answer options are given), several dummy variables have to be formed. However, a so-called *reference or base category* has to be defined. By the way, variables with an ordinal level of measurement can also be easily taken into account in the analyses.

In our data, we form four categories for religious affiliation: "Protestant", "Catholic", "other religious affiliation" and "no religious affiliation". In this case, it is sufficient to form three dummy variables and then consider the remaining fourth category as the reference category.[11] In the following example (see Table 6.10), the category "Protestant" should serve as the reference size. In the

---

[11] While the choice of the reference category for dichotomous variables is ultimately arbitrary and has no effects on the results and their interpretation, this is not always easy for qualitative variables with several categories: For example, if one takes the category 'no religious affiliation' as the reference size, then the differences between the individual denominations cannot be tested in simple analyses (however, there are tests for this). The respective reference category should also not represent the least populated category, as otherwise problems regarding the quality of the estimation, more precisely the multicollinearity (see below) can occur.

Table 6.10 Determinants of the traditional gender role orientation. (Source: GGSS 2016, $n = 1{,}458$)	Coefficient	S.E.	t-Value
Intercept	1.962	0.035	56.17
Effect of age	0.007	0.001	7.03
Effect of father's ISEI	−0.005	0.001	−6.01
Gender: female	−0.228	0.034	−6.68
Catholic affiliation	0.083	0.045	1.82
Other affiliation	0.555	0.084	6.61
No religious affiliation	−0.203	0.041	−4.96

analysis, the three dummy variables "Catholic", "other religious affiliation" and "no religious affiliation" as well as the variable gender by the dummy variable "woman" and finally the metric variables age and the ISEI of the father discussed above are taken into account.

How is such a table to be interpreted? First of all, it should be noted that all the variables considered in this model obviously have an independent influence on the traditional gender role orientation. All effects are significant ($p < 0.001$), except for that of the Catholic religious affiliation, which has a p-value of $p = 0.069$ and would only be significant at a pre-selected significance level of 10%.[12] It should be pointed out again that this is a multivariate model. In terms of content, this means that the influence of a certain variable is determined *under the control* of all the other variables in the model. Women do not have a more modern gender role orientation than men—they do, as the negative effect proves— because they are more likely to be without religious affiliation, but: Under the control of religious affiliation (and the other factors considered in the model), the traditional gender role orientation of women on a scale from 1 to 4 is 0.228 units lower than the traditional gender role orientation of men. Furthermore: Members of other denominations have a value of 0.555 higher traditional gender role orientation than Protestants, while the traditional gender role orientation of people without religion is 0.203 lower than that of Protestants. With a step on the scale of age, the traditional gender role orientation also increases by 0.007, with a stronger expression of the ISEI of the father it decreases by 0.005 points—again

---

[12] The significance can be estimated from Table 6.10 using the rule of thumb $|t| > 2$; exact p-values must be taken from the output of a data analysis program.

Table 6.11 Determinants of the traditional gender role orientation (β-coefficients). (Source: GGSS 2016, $n = 1,458$)

	β-coefficients
Effect of age	0.175***
Effect of father's ISEI	−0.149***
Gender: female	−0.163***
Catholic religious affiliation	0.052#
Other religious affiliation	0.170***
No religious affiliation	−0.142***

$\#p \leq 0.10;$ * $p \leq 0.05;$ ** $p \leq 0.01;$ *** $p \leq 0.001$

under the control of the other variables. The (corrected) variance explanation is $R^2 = 0.1423$ with the model, i. e. the independent variables explain together 14.23% of the variance of the traditional gender role orientation.

Often, an alternative representation of the results just described is also found, in which instead of the b-coefficients and the standard errors, simply the standardized values and their significance level are given, whereby the differences in the significance level are marked by a different number of stars. In Table 6.11 the corresponding results are given, whereby of course the intercept is omitted.

If one now looks at the Table 6.11 and compares the β-coefficients with each other, one will conclude that age has the greatest influence on the traditional gender role orientation and that the influence of gender and the other religious affiliation are similarly strong.

Such a comparison of the β-coefficients in multivariate analyses is quite common, but a problem should be pointed out. The β-coefficients are calculated from the b-values and the standard deviations of the independent and the dependent variables and can be interpreted as standardized changes of the dependent variables when the independent variables change by one standard deviation. If one now considers how the standard deviation is determined (see above Chap. 4), it quickly becomes clear that the independent dummy variables must always be smaller than 1, but the values of the dummy variables can only take the values 0 or 1. For dummy variables, one should therefore be careful with the interpretation of the standardized coefficients and sometimes the output of the non-standardized coefficients is also recommended in this case.

IKB—Are the prerequisites for a multivariate regression met at all? A somewhat longer excursion Often one finds—especially in older textbooks on questions of social science data analysis—at the very beginning of the presentations more or less detailed discussions about the correct prerequisites to apply the corresponding methods. This approach has its good reasons, because the methods of multivariate regression analysis described so far only provide reliable and meaningful results if a number of conditions are met. These assumptions are often also referred to as BLUE assumptions, where BLUE is an acronym for **B**est **L**inear **U**nbiased **E**stimator. Only if certain assumptions are met, one can assume that the obtained results are meaningful and not systematically biased (see in the following Urban and Mayerl 2008, pp. 177 ff.; Ohr 2010). In total, there are five different conditions:

* The relationship between the independent variables and the dependent variable is linear and not, for example, u-shaped or bell-shaped.
* The entire model is not misspecified.
* The independent and the dependent variable are measured correctly.
* The error terms of the regression follow certain rules: They do not show any discernible pattern, i.e. they are homoscedastic (Sect. 4.1). The residuals are independent of each other, they do not depend on the error term of the adjacent cases (Sect. 4.2).
* In multivariate models, the independent variables do not show too much dependence among each other, i.e. the multicollinearity must be relatively low.

In the following, these individual conditions shall be briefly examined a little more closely, without going into the statistical foundations (see, for example, Urban and Mayerl 2008).

(ad 1): Here we shall start with the linearity assumption. Especially in the social sciences, it seems a daring assumption that linear relationships exist between variables. For example, it may be that with age at marriage, the matching of partners improves, as more time was spent searching on the marriage market (see Hill and Kopp 2013). However, at a certain point, one can assume that the chances on the marriage market and thus the level of aspiration and finally the matching decrease. In the literature, therefore, an inverted u-shaped relationship between the age at marriage and the matching quality is assumed. Also with regard to the above-examined influencing factors on the traditional gender role orientation, it would be conceivable, for example, instead of a linear age effect, that the age effect increases more than linearly in higher age.

To test the assumption, the relationships between the variables are shown in simple scatter plots. However, especially in multivariate models, certain relationships can be hidden or exaggerated (see Diaz-Bone 2006, pp. 200 ff. and Chap. 7), so the partial scatter plots should be considered. In SPSS, these plots can be generated, for example, by the simple subcommand "/partialplot all". Especially in large data sets, the resulting graphics are not always informative. It is then certainly useful to perform the analyses for smaller subsamples. Whether linear relationships can be found in the (partialized) scatter plots is often a matter of interpretation. However, gross violations of the linearity rule should be recognizable in this way (see also Urban and Mayerl 2008, pp. 202 ff. for further criteria).

Naturally, the question arises in a second step how to deal with possible violations of this assumption. There are various possibilities for this: First, for example, additional terms can be included in the regression equation, such as a variable that was formed as age squared, to capture these nonlinear relationships. Second, of course, there is the possibility of including different dummy variables, in order to capture the different effects of the individual age groups.

(ad 2): It was formulated above that the entire model must not be misspecified. In detail, this can occur mainly by the "non-consideration of one or more important independent variables" or "by the consideration of one or more irrelevant independent variables" (Urban and Mayerl 2008, p. 218). However, this condition is very difficult to check, as it involves substantive and theoretical questions. In a first step, the quality criteria of the overall model, such as the F-test or the R^2-coefficient, should be considered. A non-significant overall model cannot be correctly specified. More difficult, however, are statements about the magnitude of the coefficient of determination: Especially in sociological studies, it is hard to give concrete threshold values. It can only be emphasized once again how important theoretical considerations and the concrete elaboration of empirically testable hypotheses are in this area. Without concrete ideas about social mechanisms of action, empirical analyses appear simply meaningless!

(ad 3) In this area, it is also necessary to point out again the concrete measurement of the individual variables. The corresponding operationalizations and scales should be clear and comprehensible. When looking at empirical studies, one should become critical if no hints can be found here.[13] If problems already arise in the measurement of the individual variables and constructs, the results can only be meaningful by chance (cf. as an introduction to this topic Schnell et al. 2011 or also Latcheva and Davidov 2014).

---

[13] Occasionally, one finds hints that corresponding analyses are not documented for reasons of space. While this may have been a valid argument a few years ago, considering the publication strategy of many journals, today there is no problem for everyone to make corresponding analyses accessible to interested readers, especially since some journals even offer corresponding forums.

(ad 4.1) In regression models, not all cases are predicted perfectly. We observe (almost) always certain errors or so-called residuals. A prerequisite for the validity of the corresponding results is now that these error terms follow certain rules. The first rule is the requirement of *homoscedasticity*. In terms of content, this means that the variance of the error terms should be constant across all values of the dependent variable. A deviation from this requirement occurs when there are significant differences in the error terms for certain ranges of the dependent variable. This situation is also referred to as *heteroscedasticity*. To test this assumption, one usually relies on graphical methods. For this purpose, a plot is created that shows the distribution of the error terms as a function of the predicted values. To obtain a clear picture, both variables are usually standardized.[14] The plot of these standardized variables can be easily requested in data analysis programs. In our example of the analysis of the traditional gender role orientation, the following picture results (Fig. 6.2):

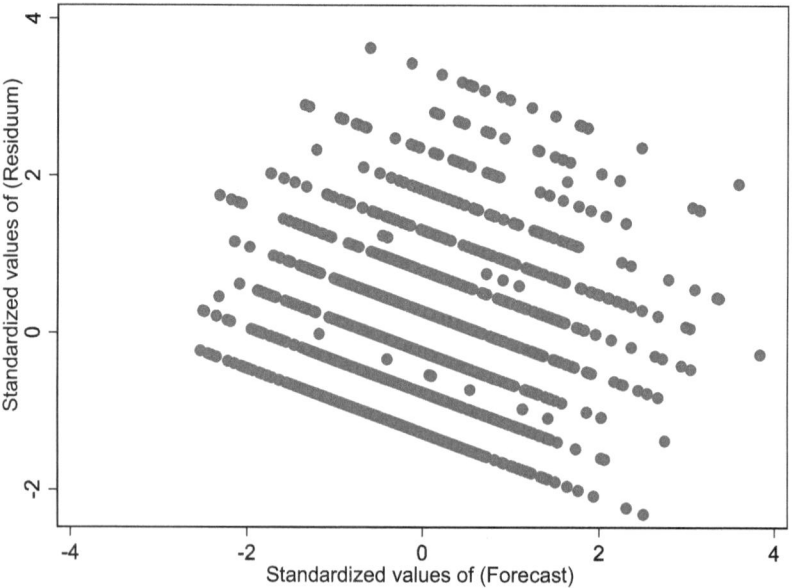

Fig. 6.2 Checking the homoscedasticity assumption. (Source: GGSS 2016)

[14]A so-called z-transformation is performed, which results in the individual variables having a mean of 0 and a standard deviation of 1.

At first glance, this scatter plot (Fig. 6.2) is difficult to interpret, as on the one hand the value range of the dependent variable is limited upwards and downwards and on the other hand the number of points is almost too large to see any patterns. Alternatively, it is possible to save the variables and look at box plots of the residuals for aggregated value ranges of the predicted values. In Fig. 6.3 there is a corresponding representation, in which four groups were distinguished.

Even when interpreting such figures, there is a certain leeway for interpretation. In this case, caution would be most appropriate with regard to the fourth group. To solve this possible problem of heteroscedasticity, the dependent variable could be transformed in a next step, so that it is as symmetrically distributed as possible (see Kohler and Kreuter 2008).

(ad 4.2) As a further condition, it was formulated above that the error terms have arisen independently of each other, i.e. they do not depend on the error term of the adjacent cases. If this is not the case, one speaks of autocorrelation. Such connections usually arise in time series and can be controlled with a simple to interpret test, the Durbin-Watson test. The range of values of this test lies between

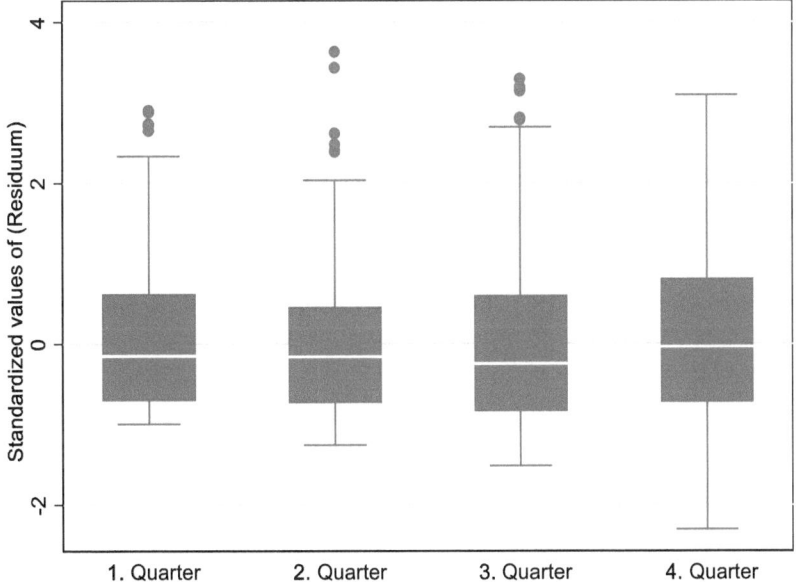

Fig. 6.3 Box plots of the standardized residuals as a function of the predicted values. (Source: GGSS 2016)

0 and 4, values close to 2 indicate a low degree of autocorrelation. However, such a test usually only makes sense for time series data, in all other cases one has to use a considerable amount of imagination to suspect order effects.

(ad 5) As the last condition, it was pointed out that the independent variables should not have too much dependence on each other. To test this, the influence of each variable on each other is examined. For this purpose, one of the independent variables is used and tested to what extent it can be explained by the other independent variables in a regression model. The coefficient of determination of this analysis is used to calculate the so-called tolerance. The better a variable can be explained by the other variables, the less independent information it provides in the original model and the more problematic the interpretation becomes. The tolerance is determined as $1 - R^2$. Tolerance values below 0.1 are considered critical, as here after all more than 90% of the variation is explained by the other explanatory variables of the original model. Instead of the tolerance, the VIF value, where VIF stands for variance inflation factor and describes very well the problem of high multicollinearity, is often given, which is determined as the reciprocal of the tolerance. Critical values of the tolerance are below 0.1 or at the VIF value above 10.

6.5   Same Results—Different Presentations

The aim of this introduction to social science data analysis is to facilitate the understanding of social science research work and to open up the possibility of critically questioning the corresponding empirical analyses.[15] If one now looks at the most important journals in the social sciences, one can be surprised by how diverse and sometimes confusing the presentation of the empirical analyses can be. This section is intended to facilitate the interpretation of these different presentations a little (cf. overall once again Allison 1999).

Already on the previous pages there are the most different possibilities to present empirical results. For example, in Table 6.10 the determinants of gender role orientation are presented by reporting the unstandardized coefficients, their stand-

---

[15]The critical readers will not have missed that the implicit assumption is that research should proceed empirically—and in a comprehensible and verifiable way. Pure theoretical considerations or contributions based on single cases certainly have their significance in the social sciences—but it should be undisputed that all hypotheses and conjectures obtained in any way have to be tested empirically. These tests are the focus here.

Table 6.12 Determinants of traditional gender role orientation. (Source: GGSS 2016, $n = 1{,}458$)

	Coefficient (Standard error)	β-coefficients
Intercept	1.962 (0.035)	
Effect of age	0.007 (0.001)	0.175^{***}
Effect of father's ISEI	-0.005 (0.001)	-0.149^{***}
Gender: female	-0.228 (0.034)	-0.163^{***}
Catholic affiliation	0.082 (0.045)	$0.052^{\#}$
Other affiliation	0.555 (0.084)	0.170^{***}
No religious affiliation	-0.203 (0.041)	-0.142^{***}

$R^2 = 0.14$
$^{\#}p \leq 0.10; \, ^{*}p \leq 0.05; \, ^{**}p \leq 0.01; \, ^{***}p \leq 0.001$

ard errors and the resulting t-values. In Table 6.11 there is an alternative presentation that only includes the standardized β-coefficients and their significance in four levels. Both forms of presentation are quite common, but not directly convertible. Instead of indicating the significance with the help of a different number of asterisks, the exact values of the significance level are often also given.

Now one can of course ask oneself why not simply all sensibly available information—unstandardized regression coefficient, standard error, standardized regression coefficient and the significance level—are given. At least two answers can be found to this question: Firstly, such a presentation is usually relatively space-consuming. (Not only) In journals, space is a scarce resource. Secondly, one can pursue quite different analysis strategies in empirical analyses (see especially Chap. 7): For example, it is often attempted to explain the bivariate influence of a variable on another quantity by further inclusion of influencing factors. To present this reasonably, a so-called gross model is usually calculated first, in which the influence of the quantity x_1 on y is shown, and then gradually additional explanatory factors x_i to x_j are added and the changes in the respective slope b_1 are considered. If one now wants to give all the above-mentioned quantities for each calculated model, one quickly reaches the limit of representability.

Even if there are certainly no very clear rules in this area, it should be borne in mind that in the unfortunately much too rarely performed replications of empirical studies, but especially in meta-analyses (see Rosenthal and DiMatteo 2001) the assessment of the results is considerably hampered if the indication of standard errors is completely omitted. Standardized effects *alone* can sometimes be difficult to interpret—especially for dummy variables. In Table 6.12 there is a suggestion to present the results already presented above in a summary way.

6.6   Afterword: a Small to-do List

We hope that our readers will study this book attentively, but it can never hurt to briefly summarize the most important steps, which can also be used as a check-list.

- Is there a clear theoretical model, whose assumptions lead to empirically testable hypotheses? If this is not the case and one moves more or less routinely in the paths of variable sociology, great caution is advised. The entire logic of empirical tests is not designed to proceed in this way and opens the door to the danger of ultimately looking at random patterns (see the very entertaining presentations by Taleb 2004).
- It is equally necessary to assume clear causal structures—we need a dependent variable. The assumed processes must be represented in linear relationships within the framework of linear regression or at least transformed accordingly.
- The primary goal is the best possible adjustment of a regression line in existing data. From the ratio of explainable variations to the total variations, a test of model fit can be determined. If this test is not significant, the entire further procedure usually makes no sense.
- By including different independent variables, one can capture the influence of a certain variable under (statistical) control of other variables.
- The influence of individual variables can be seen in the slopes, the unstandardized coefficients and in the standardized or β-coefficients, whereby the latter are not always meaningful to interpret for dummy variables.
- For the influence of individual variables, significance tests can be determined. The results of these significance tests should be taken seriously, if one has started to engage in this logic.
- Even with regression analyses, one should be critical with one's analyses. Regressions are based on a number of assumptions, which can be tested.

Finally, as further literature we recommend Arkes (2023), Gordon (2020) and Cohen et al. (2003).

References

Allison, Paul D. 1999. *Multiple regression. A primer.* Thousand Oaks: Pine Forge.

Arkes, Jeremy. 2023. *Regression analysis: a practical introduction.* London: Routledge. https://doi.org/10.4324/9781003285007.

Brüderl, Josef. 2010. Kausalanalyse mit Paneldaten. In *Handbuch der sozialwissenschaftlichen Datenanalyse*, Eds. Christof Wolf, and Henning Best, 963–994. Wiesbaden: VS Verlag. https://doi.org/10.1007/978-3-531-92038-2_36.

Bunge, Mario. 2010. Soziale Mechanismen und mechanistische Erklärungen. *Berliner Journal für Soziologie* 20:371–381. https://doi.org/10.1007/s11609-010-0130-z.

Cohen, Jacob, Patricia Cohen, Steven G. West, and Leona S Aiken. 2003. *Applied Multiple Regression—Correlation Analysis for the Behavioral Sciences.* 3. ed. London: Erlbaum Tayor & Francis, Routledge.

Diaz-Bone, Rainer. 2006. *Statistik für Soziologen.* Konstanz: UVK.

Esser, Hartmut. 1987. Warum die Routine nicht weiterhilft. Überlegungen zur Kritik an der „Variablen-Soziologie". In *Problemlösungsoperator Sozialwissenschaften*, Ed. Norbert Müller, 230–245. Stuttgart: Enke.

Gangl, Markus. 2010. Causal inference in sociological research. *Annual Review of Sociology* 36:21–48. https://doi.org/10.1146/annurev.soc.012809.102702.

Gordon, Rachel A. 2020. *Regression Analysis.* London: SAGE Publications. https://doi.org/10.4135/9781526421036878175.

Hedström, Peter, and Petri Yikoski. 2010. Causal mechanisms in the social sciences. *Annual Review of Sociology* 36:49–68. https://doi.org/10.1146/annurev.soc.012809.102632.

Hill, Paul B. 2002. *Rational-choice-theorie.* Bielefeld: Transscript. https://doi.org/10.14361/9783839400302.

Hill, Paul B., and Johannes Kopp. 2013. *Familiensoziologie. Grundlagen und theoretische Perspektiven.* Wiesbaden: Springer VS. https://doi.org/10.1007/978-3-531-94269-8_2.

Kohler, Ulrich, and Frauke Kreuter. 2008. *Datenanalyse mit Stata: Allgemeine Konzepte der Datenanalyse und ihre praktische Anwendung.* Wien: Oldenbourg. https://doi.org/10.1515/9783110469509.

Latcheva, Rossalina, and Eldad Davidov. 2014. Skalen und Indizes. In *Handbuch Methoden der empirischen Sozialforschung*, Eds. Nina Baur, and Jörg Blasius, 745–756. Wiesbaden: VS Verlag. https://doi.org/10.1007/978-3-531-18939-0_55.

Ohr, Dieter. 2010. Lineare Regression: Modellannahmen und Regressionsdiagnostik. In *Handbuch der sozialwissenschaftlichen Datenanalyse*, Eds. Christof Wolf, and Henning Best, 639–675. Wiesbaden: VS Verlag. https://doi.org/10.1007/978-3-531-92038-2_25.

Opp, Karl-Dieter. 2010. Kausalität als Gegenstand der Sozialwissenschaften und der multivariaten Statistik. In *Handbuch der sozialwissenschaftlichen Datenanalyse*, Eds. Christof Wolf, and Henning Best, 9–38. Wiesbaden: VS Verlag. https://doi.org/10.1007/978-3-531-92038-2_2.

Rosenthal, Robert, and M. Robin DiMatteo. 2001. Meta-analysis: Recent developments in quantitative methods for literature reviews. *Annual Review of Psychology* 52:59–82. https://doi.org/10.1146/annurev.psych.52.1.59.

Schnell, Rainer, B. Hill. Paul, and Elke Esser. 2011. *Methoden der empirischen Sozialforschung.* München: Oldenbourg-Verlag.

Taleb, Nassim Nicholas. 2004. *Fooled by randomness. The hidden role of chance in life and in the markets*. New York: Random House.

Urban, Dieter, and Jochen Mayerl. 2008. *Regressionsanalyse: Theorie, Technik und Anwendung*. Wiesbaden: VS Verlag.

On the Logic of Data Analysis: Which Evaluation Strategy Fits Best to My Research Question?

Before conducting an empirical research work, various conceptual considerations should take place. First of all, it is of course necessary to clarify which dependent variable should be evaluated in the analyses and which data set is available for this purpose. Which analysis methods are suitable depend on the character of the data, for example whether it is a cross-sectional or longitudinal design, and also on the level of measurement of the variables.

Another decision, which is often given too little attention, is to be made with regard to the evaluation strategy. How should one proceed when calculating a regression model? For example, should the individual independent variables be added to the model step by step or rather all in one step? When is it useful to consider interaction effects in a multiple regression model? These questions about the 'logic' of data analysis are of fundamental importance, as they concern all multivariate regression methods, that is, not only simple regression models in cross-section, but also more complex methods such as event data or panel analysis.

The choice of an evaluation strategy is based on the theoretical orientation of the research work, whereby different ideal-typical constellations can be distinguished:

- One form is characterized by the fact that the independent variables are rather equally weighted alongside each other. That is, none of the explanatory factors is theoretically or empirically particularly emphasized. For example, can differences in income between participants and non-participants in vocational training still be observed when sociodemographic variables such as age, gender and educational level are controlled for?
- Other questions aim at explaining the effect of a certain independent variable, which is in the focus of interest, by confounding variables. Typical examples

for this are internationally comparative studies or generally questions about group differences, such as between men and women or West and East Germans. To what extent can East-West differences in attitudes towards immigrants be explained by the frequency of contact that the respondent has with people with a migration background?

- For a third type of question, it is of interest to what extent the strength of the relationship between an independent variable and a variable to be explained varies depending on a confoundingvariable. For example, is the positive relationship between social support and health status stronger in developing countries than in Western industrialized nations?

In the following chapter, three evaluation strategies will be discussed and each presented using an example. The purpose for the user is to weigh up which approach best suits his or her question. Also for the recipient, the knowledge of different evaluation strategies can be useful to develop a deeper understanding of empirical analyses.

7.1   The Empirical Example

Before the different evaluation strategies are presented, a brief introduction to the empirical example on the topic of gender role orientations follows. These can generally be defined as norms of an individual about gender-specific appropriate behavior (Krampen 1979). If these norms result in an unequal status of men and women, one can also speak of sexism (Mays 2012). A classic form of sexism, which will be examined more closely here, are traditional views that assign the woman the role of wife, mother and career helper for the man.

The exact wording of the three items used, which are combined into a scale by averaging, is shown in Table 7.1. Here you can also find explanations for the operationalization of the independent variables, unless they are self-explanatory. All data transformations and analyses are also documented in the accompanying syntax for the book. To keep the example manageable, the evaluations in this chapter are limited to the GGSS 2016. A current trend analysis of the development of gender role orientations in Germany in the period 1982–2016 can be found in Lois (2020).

Before we now dive into the different evaluation strategies, we want to clarify theoretically in advance which relationships the different independent variables and the gender role orientation of a person should have. The following presentation is largely based on the study by Mays (2012).

Table 7.1 Operationalizations for the analysis of gender role orientations. (Source: GGSS 2016; Mays 2012)

Construct	Operationalization
Traditional gender role orientation (scale with value range 1–4, higher = more traditional)	"It is much better for everyone concerned if the man goes out to work and the woman stays at home and looks after the house and children."
	"It's more important for a wife to help her husband with his career than to pursue her own career."
	"A married woman should not work if there are not enough jobs to go round and her husband is also in a position to support the family." (Answer format for each item from 1 = *completely disagree* to 4 = *completely agree*)
Educational level	Measured in years of education with a value range between 8 years (no degree) and 18 years (university degree)
Church attendance	Converted into (partly estimated) church visits per year with the categories *never* = 0, *rarely* = 1, *several times a year* = 3, *1–3 × per month* = 24, *1 × per week or more often* = 52
Socialization in former GDR	Based on the federal state in which the respondent lived in his or her youth
Anomia (scale with value range 0–1 as mean of both items)	"Given the way the future looks, it can hardly be justified to bring children into the world."
	"Most politicians don't really care about the problems of ordinary people." (Response format for both items 0 = *disagree*, 1 = *agree*)
Subjective economic deprivation (scale as mean of both z-standardized items)	"And your own economic situation today?" (Response format 1 = *very good*, 2 = *good*, 3 = *partly good/partly bad*, 4 = *bad*, 5 = *very bad*)
	"Compared to how other people live here in Germany: Do you think you get your fair share, more than your fair share, somewhat less or much less?"

It is to be expected that people who grew up and were socialized in their youth in the former GDR (East Germany) have a more liberal gender role orientation than respondents who were socialized in West Germany *(socialization location hypothesis)*. This can be justified by the assumption that the egalitarian state ideology of the GDR contributed to more liberal attitudes. Another factor is the high female and especially maternal employment rate of the GDR, which was strongly promoted by the labor market and family policy. In addition, the advanced secularization in the new federal states could have contributed to comparatively liberal gender role orientations.

With regard to the educational level of a person, it can be assumed that highly educated people are more liberal-minded *(education hypothesis)*. For example, it is generally assumed that people with high cognitive abilities and knowledge stocks are more likely to question stereotypical gender roles critically. Especially for women, it should also apply that with increasing education and the associated career aspirations, the interest in egalitarian gender roles increases. Simply put: Whoever wants to make a career as a woman does not want to be disadvantaged because of gender.

Furthermore, the perception of an economic disadvantage (economic deprivation) and a social disintegration in interaction with a world perceived as insecure and threatening (anomie) can lead to insecurity (cf. Mays 2012). When people feel insecure, the further argument goes, they have an increased need for security and order. This need for security could now foster the desire for a clear social role allocation according to a traditional pattern *(deprivation hypothesis* and *anomie hypothesis)*.

Ideologically, inequalities between men and women are often legitimized in traditional religions in the name of God. In the literature, this is expressed, for example, by a hierarchical representation of the relationship between men and women. A practical-institutional example is the prohibition of ordination of women in the Catholic Church (cf. Mays 2012). Therefore, it is to be expected that religious people are more traditional-minded than secular ones *(religiosity hypothesis)*.

7.2   The Gross-Net Model

In multivariate data analysis, it is not only of interest how the individual independent variables are related to the dependent variable. In addition, the correlations between the independent variables must also be taken into account. Although it is theoretically conceivable that the effects of the independent variables are *independent*. This means that each explanatory factor influences the dependent variable independently, without—in the extreme case—being related to the other variables in the model. In practice, however, such independent effects are rather rare, as even simple sociodemographic indicators such as gender, age, education and employment status are correlated with each other.

Now the question arises, which evaluation strategies are suitable to reveal (also) the relationships between the independent variables. If the researcher does not want to emphasize any of the independent variables for theoretical reasons, the so-called *gross-net model* is suitable, which is occasionally encountered in sociological journals. The evaluation strategy here consists of determining for each independent variable first the bivariate effect, that is, without controlling for third variables, on the dependent variable (gross effect) and additionally calculating a single multiple regression model in which all relevant independent variables are included together (net effect). A comparison of gross and net effects allows a first assessment of the extent to which the respective independent variable is correlated with the other model variables.

We now demonstrate the gross-net model using our example. As discussed in Sect. 7.1, the dependent variable is a scale of traditional gender role orientations. According to the hypotheses formulated in the previous section, we can expect that the following groups of people have less traditional, i.e. rather liberal, gender role orientations: highly educated, respondents who grew up in the former GDR, less religious people, and respondents who do not feel either anomic or economically deprived.

Basically, it should be noted first that the number of cases of the bivariate linear regression models, in which we determine the gross effects of the individual variables on the gender role orientation, does not differ from the multivariate model, in which all independent variables are included together. If differences between the gross and net effects were to be observed, but the bivariate models were based on a larger number of cases than the multivariate model, this would lead to interpretation problems. In this case, there are two explanations for the observed differences: First, they can be attributed to relationships that the independent variables have with each other, and second, to differences in the number of cases caused by missing values. Such ambiguities should be avoided from the start. All models should be based on the same n. [1]

---

[1] To be able to compare gross and net effects, a so-called listwise case exclusion can be performed for both models. This means that a case is excluded from the calculation of the regression models as soon as a missing or invalid value is present for the dependent or one of the independent variables. The listwise case exclusion can be associated with a strong reduction of the initial sample in multivariate models, which reduces the chance of finding empirically existing relationships. In addition, the listwise case exclusion can lead to biased estimation results if the absence of a value is not determined by chance alone. For these reasons, so-called imputation methods have been developed, with which missing values can be replaced. An overview of these methods can be found, for example, in Allison (2005).

Table 7.2 Gross-net model for testing the relationship between sociodemographic characteristics and a traditional gender role orientation (linear regression, unstandardized b coefficients). (Source: GGSS 2016)

	Men		Women	
	Gross	Net	Gross	Net
Socialization in former GDR	−0.14**	−0.18**	−0.13**	−0.06
Economic deprivation	0.09**	0.02	0.10**	0.06*
Effects of educational level	−0.08**	−0.06**	−0.09**	−0.07**
Church attendance	0.008**	0.009**	0.013**	0.013**
Anomia	0.38**	0.29**	0.22**	0.07
Corrected r^2	–	0.12	–	0.18
n	759		748	

$+p \leq 0.10$; $*p \leq 0.05$; $**p \leq 0.01$

Table 7.2 shows the complete gross-net model. Since the effects of the different independent variables may differ by gender (see Sect. 7.3), separate models are calculated for men and women.

A content-based look at the gross effects shows that our assumptions are largely confirmed. Both men and women who were socialized in the former GDR have more liberal gender role orientations. The same applies to higher educated people. As expected, people with higher church attendance frequency and respondents who perceive increased anomie or economic deprivation are more traditionally oriented.

The gross-net model is first of all well suited to give us a general overview. If the gross effects and the net effects are relatively similar, the respective model variables have rather *independent* effects on the dependent variable, i.e. they are not or only slightly related to each other. In the example, this applies to the church attendance frequency. Proving the independence of effects is very important in many research situations. One wants to show that the influence that one or more variables have on a dependent variable is not a spurious correlation caused by other third variables.

In other places, the net effect is significantly smaller than the gross effect. This is an indication that there must be *correlations between the independent variables*. For example, while the gross effect of anomie for women is 0.22 and highly significant, the net effect is much smaller and insignificant at 0.07. Therefore, one

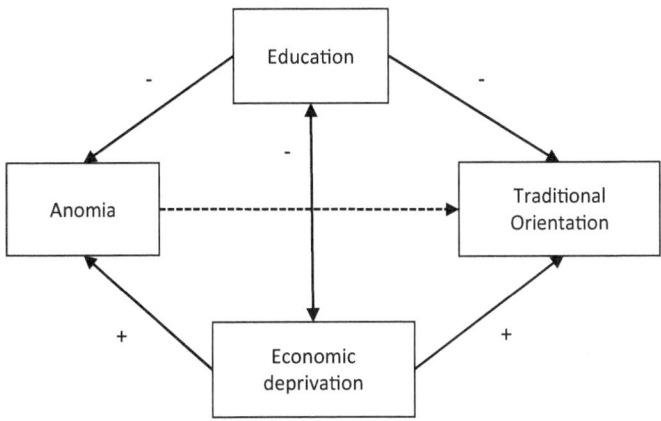

Fig. 7.1 Spurious correlation of anomie. (Own representation)

can conclude that the gross effect of anomie is a spurious correlation. How this comes about is shown in Fig. 7.1:

A correlation matrix between all variables shown in Table 7.2 (not shown) reveals that anomia correlates in the medium range with education and economic deprivation. In addition, education and deprivation also have independent effects on the dependent variable in the net model. These two characteristics are therefore third variables that cause the spurious correlation of anomia. Education correlates negatively with anomia and *simultaneously* negatively with traditional orientations, the dependent variable. Economic deprivation correlates positively with anomia and simultaneously positively with traditional gender role orientations.[2] If the characteristics education and deprivation are not statistically controlled in the gross model, the spurious effect of anomia on the traditional orientations (0.22) appears, conditioned by this correlation pattern. When education and deprivation are held constant in the net model, the anomia effect becomes insignificant. It is therefore drawn dashed in Fig. 7.1. *Independently* of education and deprivation, anomia has no independent effect.

If we return to Table 7.2, we notice another constellation: It is possible that the net effect of a variable *increases* compared to the gross effect. This is clearly shown by the socialization location of men. We will discuss this seemingly sur-

---

[2] In addition, education and deprivation correlate negatively with each other.

prising constellation, which also indicates relationships between the socialization location and other model variables, in detail in Sect. 7.2.

Even though the gross-net model, in the sense described above, provides a good general overview, it should not be overstretched in the interpretation. For example, one could focus on the question of why women socialized in East Germany are more liberally oriented than women socialized in West Germany. Is this really due to the egalitarian state ideology of the GDR? Alternative explanations would be that women socialized in East Germany are more educated and less religious.

A comparison of the corresponding gross effect for the socialization location of East German women ($b = -0.13$) with the net effect ($b = -0.06$) shows that the gross effect has clearly weakened and also become insignificant. Consequently, no independent socialization effects can be demonstrated when education, religiosity, deprivation and anomia are held constant. Consequently, the place of upbringing is related to other model variables. For example, it would be expected that a GDR socialization *manifests* itself in a low religiosity and we therefore start to argue with chains of explanation (GDR socialization $\rightarrow$ low religiosity $\rightarrow$ liberal attitude).

At this point, however, a problem of the gross-net model becomes apparent. If our focus is actually on chains of explanation, i.e. we pick out a variable such as GDR socialization, which we put at the beginning of such a chain, whose influence we want to explain by further characteristics, we quickly reach limits. It remains completely unclear *which characteristics to what extent* contribute to the fact that the gross effect of the socialization location for women (-0.13) has reduced to -0.06. Is this due to education, religiosity, economic deprivation or anomia? Put bluntly, the gross-net model here only allows us to conclude that the socialization location must somehow be related to some other variables contained in the model—an answer that is certainly unsatisfactory in this context.

One can now either accept this imprecision and simply state in general that the influence of the place of socialization is reduced when controlling for the other model variables. Such statements are quite common in sociological articles. On the other hand, one can find statements of this kind unsatisfactory and ask the question, with which independent variables the place of upbringing is exactly related, what signs these correlations have and how it is explainable in detail that the net effect has weakened compared to the gross effect. This *specific* question cannot be satisfactorily answered with the gross-net model and we will therefore introduce an alternative evaluation strategy with the mediation analysis in the next chapter.

To summarize: If the goal is to examine a single dependent variable for how it is related to several independent variables, i.e. to perform a multivariate data analysis, the gross-net model is a first evaluation strategy. It is especially applicable when the individual explanatory factors are *equally weighted* alongside each other, i.e. when the research interest does not focus on a specific independent variable.

The strategy consists of first determining the bivariate (gross) effect for each independent variable by calculating a series of bivariate regression models, each of which includes only one independent variable. Subsequently, a multiple regression model is specified, in which all independent variables enter together, to determine the net effects. A gross-net comparison can now provide first indications of whether the influences of the model variables are rather independent or whether they have strong relationships with each other and thus overlap in explaining the dependent variable.

A disadvantage of the gross-net model is that it is less well suited to statistically explain the influence of an independent variable that is at the beginning of an explanatory chain, by several other factors with which this independent variable is correlated. For questions of this kind, the mediation analysis is better suited, which is presented in the following chapter.

7.3 The Mediation Analysis

In sociology, questions are often dealt with that require a so-called mediation analysis. Generally, in these works, an independent variable is of particular interest and the analysis aims to explain the effect of this characteristic by other variables. Typical examples are internationally comparative studies, in which country-specific differences with regard to a dependent variable are first determined and then an attempt is made to statistically explain these differences by controlling for, for example, socio-structural indicators.

The so-called mediation analysis aims in general to explain the effect of an independent variable X by the fact that X influences a second, intervening independent variable Z (the so-called mediator), and Z in turn itself is related to the dependent variable Y (see Fig. 7.2): X can thus influence Y both directly and indirectly via Z. An evaluation strategy can consist of observing how the bivariate effect of X on Y changes when Z is introduced into the multiple regression model. If it becomes *smaller,* but remains significant, one speaks of a *partial mediation,* if it becomes insignificant, it is a *complete mediation.* In the latter

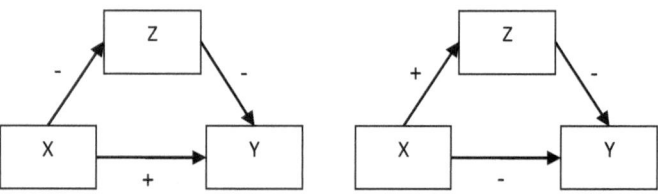

Fig. 7.2 Examples of mediation. (Own representation)

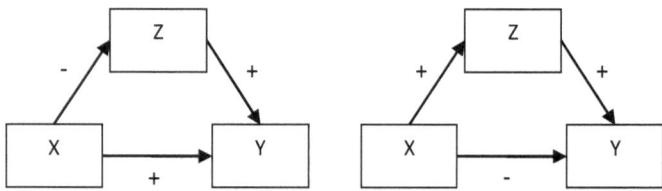

Fig. 7.3 Examples of suppression. (Own representation)

case, the direct effect of X on Y is not causal, since it can be explained at least statistically completely by the so-called intervening variable Z, the mediator.[3]

In the left example in Fig. 7.2, X has a positive direct effect on the dependent variable Y. The mediation consists in the fact that X indirectly affects Y via Z. It is now important how the sign of this indirect effect (X → Z × Z → Y) turns out overall. Here the known rules of calculation apply: Minus times plus (or plus times minus) equals minus, minus times minus equals plus and plus times plus also equals plus. The indirect effect is thus positive in the left example (minus times minus = plus) and thus has the same sign as the direct effect of X on Y. At this point we have already reached the central rule: It is then a mediation, when the direct effect of X on Y has *the same sign* as the indirect effect X → Z → Y.

The right example in Fig. 7.2 also represents a mediation accordingly. The direct effect of X on Y is negative and the indirect effect via Z as well (plus times minus = minus).

---

[3]A related term in this context is the spurious correlation. The difference between a spurious correlation and a complete mediation is that in the case of a spurious correlation Z affects X and Y, while in the case of mediation X affects Z. With cross-sectional data, which underlie our example, the actual causal direction cannot be clarified. The assumed directions of effects thus remain left to the theoretical argumentation.

If the direct and the indirect effect have *opposite* signs, it is a *suppression*. In Fig. 7.3 two examples are shown for this.

On the left, the direct effect of X on Y is positive, the indirect effect X→Z→Y is negative (minus times plus =minus). Empirically, suppression manifests itself in the fact that the bivariate effect of X on Y, when controlling for Z (the so-called suppressor), does not become weaker as in mediation, but *stronger*.[4] We observed this in connection with the gross-net model for the man's socialization location (see Table 7.2). The net effect of X, when controlling for Z, thus shows a stronger correlation with Y than the gross effect of X, that is, without controlling for Z.

Suppressions are sometimes perceived as ,disturbing' in empirical analyses, because one is usually interested in explaining a bivariate effect of X by the confoundingvariable Z, so that the X effect tends to zero more and more, until it becomes insignificant. In principle, however, suppressions can occur just as frequently as mediations and are also not less interesting.

We now want to illustrate these phenomena using our empirical example. For this purpose, we pick out a variable where there are larger deviations between the gross and the net effect. This is the case for the woman's socialization location, whose effect is known to decrease from -0.13 to -0.06 (see Sect. 7.2 and Table 7.2). Apparently, there are mediation relationships with third variables Z here, which we now want to analyze in detail.

We were therefore interested in this question: Why are women who were socialized in the former GDR more liberal than women socialized in West Germany? To stay in the terminology of path diagrams, the GDR socialization (with the reference category FRG socialization) is thus our X.

Now, first of all, it has to be clarified content-wise which intervening variables Z have to be considered. For this, we should supplement our general hypotheses from Sect. 7.1 with specific mediation hypotheses. With regard to the level of education, it is regularly found that especially the older population in the new federal states is more qualified, i. e. has more university degrees, for example, than comparable groups in the West (Statistisches Bundesamt 2020). This is attributed to the fact that many of the older people today were able to obtain high-quality educational degrees in the former GDR, for example in the technical college system. Assuming that higher education leads to more liberal gender

---

[4] ,Becoming smaller' or ,becoming weaker' always means that the respective effect moves closer to zero. ,Becoming larger' or ,becoming stronger' means accordingly that the effect moves away from zero in absolute terms.

role orientations, this results in the *education mediation hypothesis:* East German socialized women have more liberal gender role orientations than West German ones because they are more educated.

Due to long-term cultural differences and the forced secularization in the former GDR, the new federal states are also much more secular than the old ones. It is therefore expected: East German socialized women have more liberal gender role orientations than West German ones because they are less religious *(religiosity mediation hypothesis).*

For the last two indicators to be considered, anomia and economic deprivation, a suppression is theoretically more likely. Due to the consequences of transformation and greater structural problems in the East German federal states, it is expected that East German socialized people feel more politically powerless and economically disadvantaged. Both factors – anomia and deprivation – should, as shown in Sect. 7.1, be associated with more traditional gender role orientations. From this, the following two hypotheses can be derived: East German socialized women have more *traditional* gender role orientations than West German ones because they feel more anomic *(anomia suppression hypothesis)* or economically deprived *(deprivation suppression hypothesis).*

Before we calculate multiple regression models, we want to perform some descriptive analyses (Table 7.3). We look at how X (socialization location) is related to the individual Z-variables (education, religiosity, etc.). This is generally advisable in the context of a mediation analysis. It turns out that women who grew up in the former GDR, compared to women socialized in the old FRG, have a higher average level of education, are significantly less religious in terms of

Table 7.3 Descriptive statistics and *t*-tests for independent samples on socio-structural differences between women with and without socialization in the former GDR. (Source: GGSS 2016)

	Socialization in former GDR		
	Yes	No	*t*-values
Educational level	13.94	13.49	−2.53*
Economic deprivation	0.16	−0.07	−3.56**
Anomia	0.55	0.49	−2.21*
Church attendance frequency	2.76	6.91	5.01**
n	262	486	

$+p \leq 0.10$; $*p \leq 0.05$; $**p \leq 0.01$

Table 7.4 Hierarchical linear regression model to explain the effect of socialization in the former GDR on traditional gender role orientation (unstandardized b-coefficients). (Source: GGSS 2016, female respondents, $n = 790$)

	Model				
	1	2	3	4	5
Socialization former GDR	−0.13***	−0.09+	−0.10*	−0.10*	−0.06
Education level		−0.09**	−0.08**	−0.08**	−0.07**
Economic deprivation			0.05+	0.05	0.06*
Anomia				0.01	0.07
Church attendance					0.013**
Constant	1.73**	2.88**	2.82**	2.81**	2.61**
Corrected r^2	0.01	0.12	0.12	0.12	0.18

$+ p \leq 0.10$; $*p \leq 0.05$; $**p \leq 0.01$

active participation, and feel more anomic and economically deprived. All of the latter differences are significant according to t-tests for independent samples.

The descriptive analyses in Table 7.3 are thus consistently in line with our mediation and suppression hypotheses, but do not yet constitute a complete hypothesis test. In terms of our path diagrams (e.g. Fig. 7.1), we have so far only tested whether there is a relationship between X (socialization location) and Z (intervening variable, e.g. education). In addition, we now have to clarify whether the respective Z affects Y (gender role orientation).

This is done within the framework of a hierarchical (also: nested or stepwise) regression model (see Table 7.4). In a first step (model 1), only X is included to determine the gross effect, that is, a dummy variable for the socialization location (1 = former GDR, 0 = old FRG). In contrast to the gross-net model, we then do not take all the other independent variables together, but stepwise *only one* independent variable into the regression model. This way, it can be tested whether there is a mediation or suppression in each case.

As a starting point, we find in model 1 the already known gross effect of the socialization location in the amount of b = −0.13. East German socialized women are more liberally minded. The order in which we now include the intervening variables Z is ultimately arbitrary and thus follows rather theoretical considerations. The existing leeway for the user here is not entirely unproblematic. For example, it can happen that the change of X when controlling for Z is influenced by the fact that before introducing Z, there are already other independ-

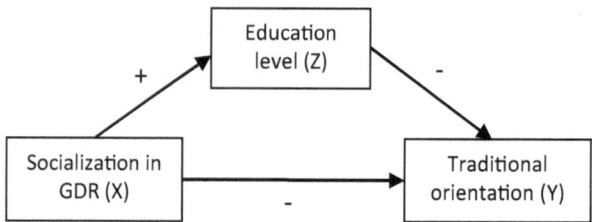

Fig. 7.4   Mediation of the socialization effect by the level of education. (Own representation)

ent variables in the model. It is therefore advisable to calculate different 'step sequences' to check the stability of the results.[5]

In model 2, the education mediation hypothesis is tested. The level of education has the expected negative effect: the higher educated a woman is, the more liberal her gender role orientation. At the same time—and this is the central aspect here—the East-West difference of -0.13 is reduced to -0.09. Since the influence of the socialization location in model 2 is still marginally significant, this is thus, in accordance with our hypothesis, a partial mediation. Let us make clear what these results mean by using a graph (Fig. 7.4)[6] :

We can read from the regression output (Table 7.4, models 1 & 2) that the effect of GDR socialization on the traditional gender role orientation is negative. We also see that the level of education has a negative influence on this orientation as well. Now it depends on the sign of the indirect effect $(X \rightarrow Z) \times (Z \rightarrow Y)$. For mediations, it is well known that this indirect effect must have the same sign as the direct effect $X \rightarrow Y$. The direct effect is negative. Moreover, we already know that it is a mediation, since the effect of the socialization location has weakened when we included the level of education. Thus, the relationship between GDR socialization and the level of education *must* be *positive*. Only in this case we cal-

---

[5]Alternatively to the procedure presented here, it is possible to remove the first intervening variable z_1 (in the example: education) from the model again before the next intervening variable z_2 (in the example: deprivation) is included, etc. However, it is still recommended to calculate at least a complete model with all independent variables (corresponding to model 5 in Table 7.4).

[6]For mediation analyses, it is actually advisable to take a sheet of paper in doubt and draw a path diagram like in Fig. 7.3 with the corresponding signs of the effects.

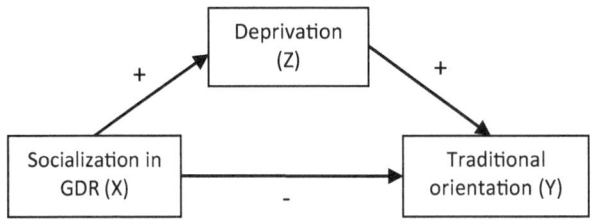

Fig. 7.5 Suppression of the place of socialization by economic deprivation. (Own representation)

culate for the indirect effect 'plus times minus = minus' and get the same sign as for the direct effect.[7]

In model 3, economic deprivation is additionally included as an intervening variable. We see that the effect of East socialization slightly *increases* from −0.09 to −0.10. It is therefore, as theoretically expected, a *suppression*. Let us again use a graphical illustration, see Fig. 7.5.

Again, we can read from the regression output (Table 7.4, Models 1 & 3) that GDR socialization has a negative and economic deprivation (at least tendentially) a positive effect on traditional orientations. Moreover, we know that it is a suppression, since the socialization effect *increased* when introducing deprivation. For suppressions, the following applies: The indirect effect has the opposite sign to the direct effect. The direct effect is negative. Thus, the sign of the indirect effect (GDR socialization→ deprivation→ traditional orientation) must be positive. The effect 'deprivation→ traditional orientation' is positive. Thus, the effect 'GDR socialization→ deprivation' *must* also be positive. Only in this case do we calculate 'plus times plus = plus'.

How is this suppression now to be interpreted or described? Technically speaking, the "part" of GDR socialization (X) that is adjusted for Z (economic deprivation) is more strongly associated with gender role orientation (Y) than before, that is, without controlling for Z. One can also imagine the following situation: If economic deprivation did not differ between women with and without GDR socialization, women socialized in the GDR would be even more liberally minded than they already are. And again, in other words: *Although* East German

---

[7] We also already know from the descriptive analysis (Table 7.3) that East German socialized women are more educated.

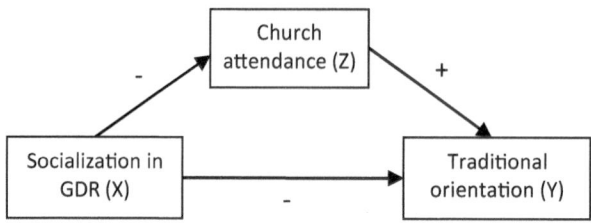

Fig. 7.6 Mediation of the socialization location by church attendance. (Own representation)

socialized women are more economically deprived and this deprivation makes their gender role orientation more traditional, they are *apart from that* more liberally minded than West German socialized women.

In model 4, we control for anomie attitudes and find that the socialization effect does not change. This is probably not due to the fact that there is no difference in anomie between East German and West German socialized women (partial effect $X \rightarrow Z$). We have shown in Table 7.3 that the perceived anomie among East German socialized women is significantly higher. However, anomie has no effect on gender role orientation in model 4. I.e., the partial effect $Z \rightarrow Y$ is not significant and we consequently reject our deprivation-suppression hypothesis.[8]

Finally, in model 5, church attendance is included. The socialization effect then decreases relatively strongly (from -0.10 to -0.06) and becomes insignificant. It is therefore, as expected, a mediation. This is, in combination with the educational level , which is still in the model, now complete. Let us also look at this mediation graphically (Fig. 7.6).

The indirect effect 'GDR socialization $\rightarrow$ church attendance $\rightarrow$ traditional orientation' is negative (minus times plus = minus) and thus has the same sign as the direct effect (GDR socialization $\rightarrow$ traditional orientation).

Let us now briefly discuss the advantages of the mediation analysis presented. Compared to the gross-net model, the step-by-step procedure showed in detail how the effect of the socialization location on the gender role orientation is explained (mediation) or concealed (suppression) by the other model variables.

---

[8] It should be noted that anomie in the specification of model 4, i.e. under control of deprivation and education level, has no effect on Y. The situation might be different if anomie were included in model 2, i.e. without controlling for other variables.

While the results in the gross-net model are not clear in this regard, the mediation analysis can thus provide a detailed picture of the correlations between the independent variables. It became clear that only *one* new independent variable should be included per regression step in order to be able to draw unambiguous conclusions about mediation, suppression or independent effects.[9]

A not to be underestimated problem in the interpretation of mediations and suppressions in conventional stepwise regression analyses is that the significance of the indirect effects cannot be assessed without additional evaluations. For example, the effect of the socialization location decreases quite clearly when controlling for church attendance (from $b = 0.10$ to $b = 0.06$, see Table 7.4, models 4 & 5). Nevertheless, this result alone does not imply that the corresponding indirect effect 'GDR socialization $\rightarrow$ church attendance $\rightarrow$ traditional orientation' is actually significant at least at the 5 % level. However, there is the possibility to answer this question with the help of the so-called Sobel test (Sobel 1982). The corresponding procedure is, using the example of the mediator church attendance, shown in the short excursion below.

Let us summarize: To determine the relationship patterns between the independent variables of a multiple regression model in more detail, a simple evaluation strategy is useful: First, the gross effect of the independent variable X, which is the focus of the content, on the dependent variable Y is determined. Then, the third variables Z are gradually introduced into the regression model, one at a time per step. From the change in the X coefficient when introducing a Z, one can now conclude how X, Z and Y are related. If the X coefficient does not change, X and Z are not significantly related and have independent effects on Y. If the X coefficient is reduced when holding a Z constant, it is a mediation. In this case, we explain through Z why X affects Y. If the X effect becomes *stronger* when controlling for Z, it is a suppression. The influence that X has on Y is not explained by Z, but on the contrary: It is concealed.

---

[9]One could object here that it is cumbersome and space-intensive in the presentation to include each independent variable individually in the regression model. However, two things have to be distinguished here: first, the analyses that the researcher performs and second, the presentation of the results. For example, it may be sufficient in some cases to present only one gross-net model in tabular form. In this case, however, at least in the text, it should be explained what causes the larger deviations between the gross and net effects in detail, i.e. where mediation and suppression relationships exist. A detailed evaluation of the data, in which the independent variables are included step by step, is therefore recommended in any case.

Table 7.5 Calculation of the partial effects for the mediation "Socialization former GDR → Church attendance frequency → Traditional gender role orientation" (linear regression, $n = 748$). (Source: GGSS 2016)

Church attendance → Gender role orientation		
b_1	SE_{b1}	t-value
0.0134	0.0018	7.37

Socialization former GDR → Church attendance		
b_2	SE_{b2}	t-value
−4.146	0.926	−4.48

A limitation of these stepwise regression models is that the change in the X coefficient alone does not allow to infer the significance of the respective indirect effects. Here, additional analyses may have to be performed, for example using the Sobel test or specialized analysis software.[10]

Excursus on the Sobel test The calculation of the Sobel test can be done 'by hand' in three steps. As an example, we take the indirect effect 'Socialization place GDR → Church attendance → Traditional orientation'. In a first step, two bivariate[11] regression models are calculated, which capture the influence of church attendance frequency on traditional gender role orientation and the influence of GDR socialization on church attendance. This results in the results shown in Table 7.5.

In a second step, the indirect effect can be calculated as the product of the two direct effects:

$$b_{\mathrm{ind}} = 0.0134 * -4.146 = -0.055$$

---

[10]The following literature on mediation and suppression is mentioned, which can help the reader to obtain further information: Classical works on mediation analysis are by Baron and Kenny (1986) and James and Brett (1984). Another application-oriented introduction can be found, using the example of linear regressions, in Urban and Mayerl (2018, Chap. 6). The path-analytic determination of indirect effects is explained by Bollen (1987). Detailed technical details and extensions to mediation analysis can be found in MacKinnon (2008).

[11]For simplification reasons, the model is specified more parsimoniously here than in Table 7.4 (Model 5) where education, anomie and deprivation are additionally controlled for.

Using the Sobel test, the significance of this indirect effect can now be determined. For this purpose, a z-value is calculated according to the following formula:

$$z = \frac{b_1 \cdot b_2}{\sqrt{\left(b_1^2 \cdot \mathrm{SE}_{b2}^2\right) + \left(b_2^2 \cdot \mathrm{SE}_{b1}^2\right)}}$$

$$= \frac{-0.055}{\sqrt{\left(0.0134^2 \cdot 0.926^2\right) + \left(-4.146^2 \cdot 0.0018^2\right)}} = -3.84$$

On this basis, the following conclusions can be drawn in the present case:

- The indirect effect 'socialization former GDR $\rightarrow$ church attendance $\rightarrow$ gender role orientation' is negative, as is the direct effect 'socialization former GDR $\rightarrow$ gender role orientation'. Since the direct and indirect effects have the same sign, this is a mediation.
- The indirect effect is highly significant with a z-value of -3.8 with one degree of freedom.[12]

However, the Sobel test in its simple form can only be applied under the conditions that the dependent variable as well as the mediator (Z) are metric and that there are no other variables in the model besides X and Z. If these conditions are not met, the significance of the indirect effects can be determined, for example, by using programs for the analysis of structural equation models. For instance, the software Mplus (Geiser 2011) has a corresponding functionality for the inferential statistical test of indirect effects.[13]

---

[12] The z-value -3.84 corresponds to the empirical significance level $p = 0.00012$ in the standard normal distribution for a two-sided alternative hypothesis. The probability of an alpha error—the indirect effect is zero in the population—is thus smaller than 0.001 %.

[13] Those who do not have access to specialized software like Mplus can either calculate by hand or use the installation of additional modules. For Stata, the additional module *mediation* is recommended (see https://econpapers.repec.org/software/bocbocode/s457294.htm) and for SPSS the macro PROCESS (see http://www.processmacro.org/download.html, both last accessed on 06.02.2021). Those who prefer to calculate by hand, but want to save some work, can use the online calculator for the Sobel test by Kristopher J. Preacher (http://quantpsy.org/sobel/sobel.htm; last accessed on 13.12.2020).

7.4   The Moderation Analysis

In the evaluation strategies discussed so far, the initial situation was always that we found a bivariate influence of at least one independent variable X on the dependent variable Y (for example, an effect of the place of socialization on the gender role orientation). Subsequently, it was checked whether this bivariate effect is independent, or whether it is a spurious correlation caused by a third variable Z. However, the third variables can have another important function, as we will see in the following: They can specify the *conditions* under which the influence of an independent variable X on a dependent variable Y is stronger or weaker.

Let us illustrate this briefly with an everyday example: As we know from painful experience, we are more often cold (and possibly also infected with COVID-19) in winter than in summer. In this example, the cold viruses are our X, which causally influences the dependent variable Y (being sick or healthy). In winter, however, the viruses have a better chance of infecting us than in summer, because we spend more time with other people in closed and heated rooms, for example. The season *moderates* the relationship between cold viruses and the risk of getting sick.

Let us illustrate the moderation[14] a bit more formally using a diagram (Fig. 7.7). The moderator Z here points to the arrow that goes from X to Y. Z thus indicates the conditions under which the strength of the relationship between

Fig. 7.7 Schematic representation of a moderation. (Own illustration)

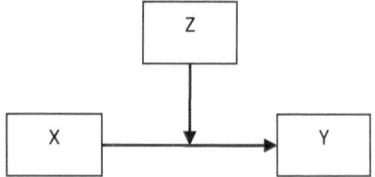

---

[14] In the course of this chapter, the terms mediation, suppression and moderation were introduced. This terminology comes from the English-language literature and is used relatively consistently there. In German-language sociological methods books, however, the terminology is partly inconsistent. Mediation is, for example, referred to by Diekmann (2010, p. 726) and Schnell et al. (2011, p. 227) as "explanation" or "interpretation" and by Kühnel and Krebs (2001, p. 473) as "confounding". Moderation is called "prediction" by Schnell et al. (2011, p. 227), while Diekmann (2010, p. 730) speaks of "specification".

X and Y varies. Unlike mediation, it is *not* necessary for the moderator Z to correlate with X or with Y. A moderation analysis is therefore, as sometimes overlooked, a completely independent evaluation strategy that has nothing to do with the phenomena discussed before, such as mediation and suppression.

Moderation analyses are often performed at an advanced stage of research. For example, suppose we have already examined in detail, using a mediation analysis, the extent to which socialization in the former GDR is related to gender role orientation. In this case, in a next step, it can be clarified whether the strength of the socialization effect varies between different groups, e.g. between men and women or between different birth cohorts. In addition, moderation analyses can also be used when, referring to *all persons* in the data set, no relationship between X and Y is shown. The goal then is to test the question of whether this effect may only occur in specific population groups, e.g. in certain social situations or milieus.

Let us now illustrate the moderation analysis using our example of gender role orientations. Here too, it is necessary to do some theoretical work and formulate some moderation hypotheses. Following the argumentation of Mays (2012), it can be expected that full-time employed women have a strong interest in egalitarian gender roles. The basic assumption is that actors are more likely to have certain attitudes if they themselves benefit from the assumed attitude object. For example, employed women could be more egalitarian because they want to avoid wage discrimination. This argumentation cannot be easily transferred to men. Employed men could have a stronger interest in a traditional division of roles, which allows them to focus more on their profession and avoid professional competition from women. Gender can therefore be regarded as a characteristic that *moderates* the relationship between full-time employment and traditional gender role orientation. We expect a negative relationship between employment and traditional attitudes for women and no or a positive relationship for men *(employment status moderation hypothesis).*

For the mother's employment, it can further be assumed that people develop a more egalitarian understanding of roles if they have experienced the model of 'employed mother' in their own family. Here too, Mays (2012) expects a gender difference: She suspects that boys are especially drawn to or have been drawn to household chores in their own family if their mother is employed. In the course of this, they develop more egalitarian ideas. They learn that household chores are not per se women's and employment men's business. For girls, on the other hand, it can often be assumed that they participate or have to participate in household chores regardless of their mother's employment. Again, gender is thus the moderator. It is expected that the negative relationship between the mother's employment and traditional gender role orientations is more pronounced for men than for women (*mother's employment moderation hypothesis*).

Furthermore, Mays (2012) finds on the basis of data from the GGSS 2008 that one's own religiosity and the religiosity of the parents play a much less important role for men's gender role orientation than for women. We want to check whether we can confirm this moderation effect also for the GGSS 2016. Since we do not yet have a well-developed theoretical argumentation for this case, we formulate a research question instead of a hypothesis: Is the relationship between individual religiosity (here: frequency of church going) and gender role orientation moderated by gender?

The first step in our moderation analysis is to calculate *separate regression models* by gender for the variables that are the subject of our moderation hypotheses. We divide the GGSS sample by gender into approximately two equal groups and calculate bivariate regression models with the independent variables own employment, mother's employment and frequency of church attendance in the subsamples of men and women (see Table 7.6).

The results are consistent with the moderation hypotheses: The influence of the mother's employment is indeed somewhat stronger for men than for female respondents. For one's own employment, on the other hand, a stronger effect is found for women, in line with the corresponding hypothesis. Moreover, we also come to the conclusion with the GGSS 2016 that the positive effect of church attendance on traditional orientations is stronger for women than for men.

The procedure of calculating separate models for the levels of a moderator is particularly suitable for categorical moderators with few levels. Separate models are certainly a good first step to investigate moderation, but they are not sufficient by themselves. At first glance, for example, the difference in the effect of one's

Table 7.6 Gross effects of various sociodemographic characteristics on gender role orientation depending on gender (linear regression, unstandardized b coefficients). (Source: GGSS 2016)

	Men	Women
Mother's employment	−0.24**	−0.18**
Full-time employment (respondent)	−0.19**	−0.33**
Church attendance	0.008**	0.013**
n	759	748

$+p \leq 0.10$; $*p \leq 0.05$; $**p \leq 0.01$

own employment between men and women (−0.19 versus −0.33) may appear relatively large. However, this does not yet allow to conclude with certainty that the effect of employment status *statistically significantly* differs between men and women.

To perform this significance test, the calculation of *interaction effects* is necessary. These are technically formed by multiplying an independent variable X with a moderator Z. The multiplicative term is then added to the main effects of X and Z in a multiple regression model. It is important to note that both the main effects and the interaction effect from these two variables are included in the model. If the main effects are not in the model, the interaction effect is not interpretable. In addition, it is occasionally useful to center metric moderators, e.g. age, before calculating the interaction effect, as this facilitates the interpretation of the conditional main effects. Centering means subtracting the arithmetic mean from a metric variable. We will come back to this topic shortly.

Let us now take a look at Table 7.7, in which the interaction effects for our research question are shown. First, one needs to know how to interpret the conditional main effects. This seems to cause difficulties frequently in moderation

Table 7.7 Moderation of effects of sociodemographic variables on traditional gender role orientation by gender (linear regression, unstandardized *b*-coefficients). (Source: GGSS 2016, $n = 1507$)

	Model		
	1	2	3
Conditional main effects			
Woman	−0.22**	−0.19**	−0.23**
Mother's employment	−0.24**		
Full-time employment (respondent)		−0.19**	
Church attendance			0.008**
Interaction effects			
Woman × Employment status of mother	0.06		
Woman × Full-time employment		−0.14*	
Woman × Church attendance			0.005+
Constant	2.01	1.98	1.84**
Adjusted r^2	0.04	0.06	0.06

$+p \leq 0.10$; $*p \leq 0.05$; $**p \leq 0.01$

analyses, as even in leading sociological journals, examples of inaccurately or often even wrongly interpreted conditional main effects can be found regularly. The rule is as follows: When one has included the interaction effect from two independent variables X and Z in a multiple regression model, the conditional main effect of X refers to the case Z = 0 and the main effect of Z applies accordingly at X = 0.

Let us go through this, together with the interaction effects, in Table 7.7 for the individual models. In model 1, the maternal employment moderation hypothesis is tested. The model contains the two main effects 'woman' (with the values 1 = woman, 0 = man) and 'mother's employment' (1 = yes, 0 = no) as well as the interaction effect 'mother's employment × woman'. The conditional main effect of 'woman' is −0.22. It can be interpreted as follows: The traditional gender role orientation is 0.22 points lower for women than for men, *if the respondent's mother is not or was not employed.* It is important to always pay attention to the italicized condition when interpreting. Since an interaction effect is included in the model, this is a conditional main effect. Accordingly, the interpretation of the influence of maternal employment is also different than usual: *Male respondents* (woman = 0) are 0.24 points less traditionally oriented if their mother is or was employed. We also find this conditional main effect in Table 7.6 (left column).

How is the interaction effect to be interpreted, which is our main focus? It expresses that the effect of maternal employment is 0.06 points more positive for women than for men. Since the effect for men is negative (−0.24), this means: The influence of maternal employment is 0.06 points weaker for women. The interaction effect tells us how large the *difference* in the effect of maternal employment between men and women is. For men, the effect is −0.24. For women, it is −0.24 + 0.06 = −0.18 (see also Table 7.6). The difference between −0.24 and −0.18 is 0.06 (= interaction effect).

Interaction effects are symmetrical and can always be interpreted from two perspectives. We can also read it as follows: The difference in the traditional gender role orientation between men and women decreases by 0.06 points if the respondent's mother is or was employed. The gender difference is −0.22 when the mother is not employed (conditional main effect). If she is employed, the difference between men and women is −0.22 + 0.06 = −0.16. This reading is less useful in light of our hypothesis, but generally just as possible.

The most important point in connection with model 1 has not been mentioned yet: The interaction effect is not significant. Therefore, we reject the maternal employment moderation hypothesis. We cannot rule out with sufficient certainty that the gender difference in the effect of maternal employment is purely coincidental. However, we only come to this conclusion when we know the significance level of the interaction effect. This shows that the calculation of these interactions in the context of moderations is ultimately indispensable to make a definitive decision about the acceptance or rejection of moderation hypotheses.

In model 2, we test the employment moderation hypothesis and start again with the two conditional main effects. Women are 0.19 points less traditionally oriented than men *if they are not full-time employed.* And the gender role orientation of full-time employed *men* is 0.19 points less traditional than that of non-full-time employed men. The interaction effect, viewed from the perspective of our hypothesis, says the following: The effect of full-time employment is 0.14 points more negative for women than for men. For men, this characteristic has an effect of -0.19 and for women of $-0.19 - 0.14 = -0.33$. Since the interaction effect is also significant, we accept the employment moderation hypothesis.

In model 3, the situation is somewhat different, as we now consider the frequency of church attendance, an ordinal characteristic with several levels. Here, too, the question arises whether the influence of this characteristic is moderated by gender. The conditional main effect of gender—women are 0.23 points less traditional—applies *at a frequency of church attendance of zero.* This case is meaningful to interpret: the respondent never attends church. For other characteristics, such as age, however, there may be problems of interpretation at this point. Let us replace, for a moment, in model 3, the frequency of church attendance with age. Then the conditional main effect of 'female' would refer to an age of zero years. At this age, however, one does not yet have a gender role orientation. Therefore, such a main effect could be perceived as unsatisfactory, as it refers to an implausible constellation. The solution here would be to center the age. To do this, simply create a new variable 'centered age' by subtracting the arithmetic mean from the age. After centering, the value zero would correspond to an *average* age, positive values would be above average, negative below average. In this way, a meaningful interpretation of the main effect can be ensured.

Let us return to the example. The conditional main effect of the frequency of church attendance means that *among men* the gender role orientation becomes 0.008 points more traditional with each annual church attendance. The interaction effect shows that the influence of the frequency of church attendance is 0.006 points more positive and thus stronger for women than for men. The interaction is at least marginally significant. *Among women* the effect of church attendance is $0.008 + 0.005 = 0.013$.

Finally, we should think briefly about the following question: Is it sensible to ignore moderation, i.e. simply calculate a common model for men and women and omit all interaction effects? In this case, we would implicitly assume by our modeling that the strength of the influence of the respective independent variables on the gender role orientation does not differ between men and women. However, this assumption is not true, as we have seen with the help of the above moderation analysis, in some cases—employment, church attendance. A model without interaction effects would therefore be strictly speaking misspecified, as it is not sufficiently adapted to the observed data.

To summarize: In contrast to mediation or suppression, we do not ask ourselves in moderation analyses whether a third variable Z either explains or suppresses the effect of an independent variable X on a dependent variable Y, because it is related to X and Y. Z rather indicates, in moderation, under which conditions the direction or strength of the effect of X on Y varies, *without* necessarily being related to X or Y. A simple evaluation strategy for analyzing moderations is to determine the effect of X on Y separately for the different levels of Z, the moderator. However, whether the relationship between X and Y is significantly moderated by Z can only be seen from the calculation of an interaction effect. This is a multiplicative term of X and Z, which, together with the main effects of X and Z, enters a multiple regression model.[15]

7.5   Afterword

Let us summarize the benefits of knowing the evaluation strategies presented. The user is thus enabled to bring the theoretical orientation of the work to the point and to choose an evaluation strategy that is tailored to this question. Should the research question, for example, be 'influences of sociodemographic characteristics on gender role orientation' or 'why are persons with GDR socialization less traditional?' In the first case, obviously several equally important explanatory factors are discussed, while in the second case the influence of a specific variable, the place of socialization, is in the focus, which can possibly be explained by further third variables. In the first case, a gross-net model would be a sensible option, in the second a detailed mediation analysis. However, the work could also have the following title: 'Sociodemographic characteristics and gender role orientation—do the effects differ between men and women?' In this study, a com-

---

[15] Also for moderation, the reader is recommended further literature. In addition to the fundamental work by Baron and Kenny (1986) on conceptual aspects, the technical details of calculating interaction effects are discussed in detail in Aiken and West (1996), Fox (1997), Frazier et al. (2004) and Whisman and McClelland (2005) and as an introduction in Urban and Mayerl (2018, Chap. 6) as well as Lohmann (2010).

pletely different goal would be pursued than in the two works mentioned above. For example, it could be checked within the framework of a moderation analysis to what extent the direction and the strength of the church attendance effect varies depending on the gender.

Restrictively, it should be noted that the evaluation strategies discussed should not necessarily be understood as 'either—or'. It can make sense to combine several analysis perspectives, such as a mediation and moderation analysis, in one work. Also, when critically receiving professional articles, the logic of data analysis should be kept in mind: Are the relationships between the independent variables discussed in the respective work? Do the author(s) expect theoretically independent effects of the independent variables or do they assume mediator, suppressor or moderator constellations? Are mediation or suppression relationships clearly recognizable in the empirical evaluations, since the independent variables are each added individually and stepwise to a regression model or corresponding explanations are found in the text? Are indirect effects tested for significance, for example with the help of the Sobel test? Is moderation assessed only on the basis of separately calculated models or additionally by calculating interaction effects? Taking into account such aspects can certainly help to sharpen the critical eye when assessing empirical analyses. For further reading, we recommend Baron and Kenny (1986), Frazier et al. (2004), James and Brett (1984), and MacKinnon (2008).

References

Aiken, Leona S., und Stephen G. West. 1996. *Multiple regression: Testing and interpreting interactions*. Newbury Park: Sage.

Allison, Paul D. 2005. *Missing data. Quantitative applications in the social sciences*. Thousand Oaks: Sage.

Baron, Reuben M., und David A. Kenny. 1986. The moderator-mediator distinction in social psychological research: Conceptual, strategic and statistical considerations. *Journal of Personality and Social Psychology* 51:1173–1182. https://doi.org/10.1037/0022-3514.51.6.1173.

Bollen, Kenneth A. 1987. Total, direct, and indirect effects in structural equation models. In *Sociological methodology,* Hrsg. Clifford C. Clogg, 37–69. Washington, D.C.: American Sociological Association. https://doi.org/10.2307/271028.

Diekmann, Andreas. 2010. *Empirische Sozialforschung. Grundlagen, Methoden, Anwendungen*. Reinbek: Rowohlt.

Fox, John. 1997. *Applied regression analysis, linear models, and related methods*. Thousand Oaks: Sage.

Frazier, Patricia A., Andrew P. Tix, und Kenneth E. Barron. 2004. Testing moderator and mediator effects in counseling psychology research. *Journal of Counseling Psychology* 51:115–134. https://doi.org/10.1037/0022-0167.51.1.115.

Geiser, Christian. 2011. *Datenanalyse mit Mplus. Eine anwendungsorientierte Einführung.* Wiesbaden: Springer VS. https://doi.org/10.1007/978-3-531-93192-0.

James, Lawrence R., und Jeanne M. Brett. 1984. Mediators, moderators, and tests for mediation. *Journal of Applied Psychology* 69:307–321. https://doi.org/10.1037/0021-9010.69.2.307.

Kühnel, Steffen-M., und Dagmar Krebs. 2001. *Statistik für die Sozialwissenschaften. Grundlagen, Methoden, Anwendungen.* Reinbek: Rowohlt. https://doi.org/10.1007/s11615-003-0126-9.

Krampen, Günter. 1979. Eine Skala zur Messung der normativen Geschlechtsrollen-Orientierung (GRO-Skala). *Zeitschrift für Soziologie* 8:254–266. https://doi.org/10.1515/zfsoz-1979-0304.

Lohmann, Henning. 2010. Nicht-Linearität und Nicht-Additivität in der multiplen Regression: Interaktionseffekte, Polynome und Splines. In *Handbuch der sozialwissenschaftlichen Datenanalyse*, Hrsg. Christoph Wolf und Henning Best, 677–707. Wiesbaden: VS Verlag. https://doi.org/10.1007/978-3-531-92038-2_26.

Lois, Daniel. 2020. Gender role attitudes in Germany, 1982–2016: An age-period-cohort (APC) analysis. *Comparative Population Studies* 45:35–64. https://doi.org/10.12765/CPoS-2020-02.

MacKinnon, David P. 2008. *Introduction to statistical mediation analysis.* Milton Park: Routledge. https://doi.org/10.4324/9780203809556.

Mays, Anja. 2012. Determinanten traditionell-sexistischer Einstellungen in Deutschland – Eine Analyse mit Allbus-Daten. *Kölner Zeitschrift für Soziologie und Sozialpsychologie* 64:277–302. https://doi.org/10.1007/s11577-012-0165-6.

Schnell, Rainer, Paul B. Hill, und Elke Esser. 2011. *Methoden der empirischen Sozialforschung.* München: Oldenbourg.

Sobel, Michael E. 1982. Asymptotic confidence intervals for indirect effects in structural equation models. In *Sociological methodology*, Hrsg. Samuel Leinhardt, 290–312. Washington, D.C.: American Sociological Association. https://doi.org/10.2307/270723.

Statistisches Bundesamt. 2020. *Internationale Bildungsindikatoren im Ländervergleich.* Ausgabe 2020 – Tabellenband. Wiesbaden.

Urban, Dieter, und Jochen Mayerl. 2018. *Angewandte Regressionsanalyse: Theorie, Technik und Praxis.* Wiesbaden: Springer VS. https://doi.org/10.1007/978-3-658-01915-0.

Whisman, Mark A., und Gary H. McClelland. 2005. Designing, testing, and interpreting interactions and moderator effects in family research. *Journal of Family Psychology* 19:111–120. https://doi.org/10.1037/0893-3200.19.1.111.

Logistic Regression

In the following chapter, the basic principles of logistic regression are explained, which, like linear regression, belongs to the standard methods of social science data analysis. Generally speaking, logistic regressions are suitable for analyzing *nominal* scaled dependent variables. Here, a distinction must be made between *dichotomous* and *multi-graded* nominal variables. If the dependent variable has only two values (i.e. is dichotomous), the *binary* logistic regression is applied, which is the focus of this chapter. If one takes a look at sociological journals, one can find various example questions for this: What are the determinants of single motherhood (Konietzka and Kreyenfeld 2005)? What does it depend on whether people participate in surveys (Koch 1994)? Which characteristics are associated with the probability of inheriting (Szydlik and Schupp 2004)? If the dependent variable has several (at least three) values, which cannot be meaningfully ordered ascending or descending, the *multinomial* logistic regression is used, which is an extension of the binary logistic regression.

In the following, it will be briefly explained why it is not appropriate to use linear regression in the case of nominal dependent variables. Subsequently, two basic concepts will be discussed, which are central for the understanding of logistic regressions: odds and probability. In the following sections, it will then be discussed in detail how these basic concepts are technically implemented in logistic regression and how to interpret the various elements of the model output.

Why is linear regression not suitable for nominal dependent variables? In Chap. 6 it was discussed that various conditions must be met for linear regression to lead to unbiased estimates; this was referred to as the so-called BLUE assumptions ("best linear unbiased estimator"). One of these conditions is the homoscedasticity of the residuals. This means for a regression with the dependent variable

F. Hartmann et al., *Social Science Data Analysis*,
https://doi.org/10.1007/978-3-658-41230-2_8

Y and the independent variable x the following: The variance of the errors, i.e. the deviations between predicted and observed values of y, must not systematically depend on the x values.

To check whether this condition holds in logistic regressions, we take a look at a scatter plot (Fig. 8.1). On the y-axis, the dependent variable is shown, which has two values (1 and 0). Accordingly, there are also only two observed values (1 or 0), which are represented by the points.

The area between 0 and 1 can be interpreted as the *probability* that the dependent variable y takes the value 1. Now one could simply calculate a linear regression on y, the corresponding regression line is drawn in the diagram. It can be seen that the probability for y = 1 increases with increasing x-values, x thus has a positive effect on y. The values lying on the regression line are the predicted values for y = 1. Since these are probabilities, this regression is called a *linear probability model.*

Here too, the residuals are defined as deviations between the predicted values (that is, the regression line) and the observed values (the points). It is immediately

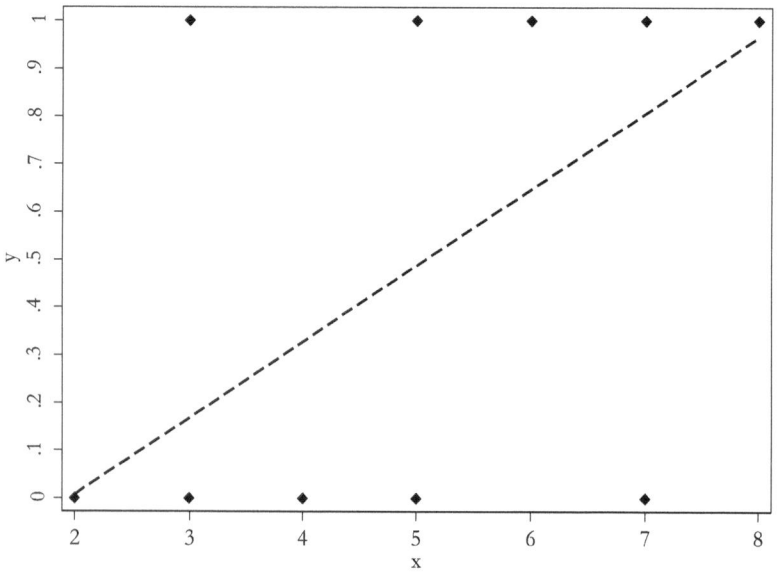

Fig. 8.1 Scatter plot for a dichotomous dependent variable Y and a metric independent variable X including regression line. (Own representation)

apparent that for very small x-values (for example x = 2) or very large values (for example x = 8) only minor deviations between line and points occur. In the middle range of values (for example at x = 5), however, larger errors are made. Since the variance of the residuals of the linear probability model is systematically related to the x-values, the residuals are *heteroscedastic*. Thus, one of the BLUE assumptions is not met.

A second problem that can arise in the context of a linear probability model is shown in Fig. 8.2. Here, two functions for the (positive) relationship between a dichotomous dependent variable y and a metric independent variable x are shown. The dashed line represents again the linear probability model and the solid line represents the so-called *logit function,* that is, a logistic regression model, which will be explained in more detail below.

The course of the linear regression function reveals the following problem: If the x-value is smaller than about −8 or larger than 8, the linear model predicts probabilities smaller than zero or larger than one, respectively. These values are not defined, however, since probabilities are by definition between 0 and 1. The

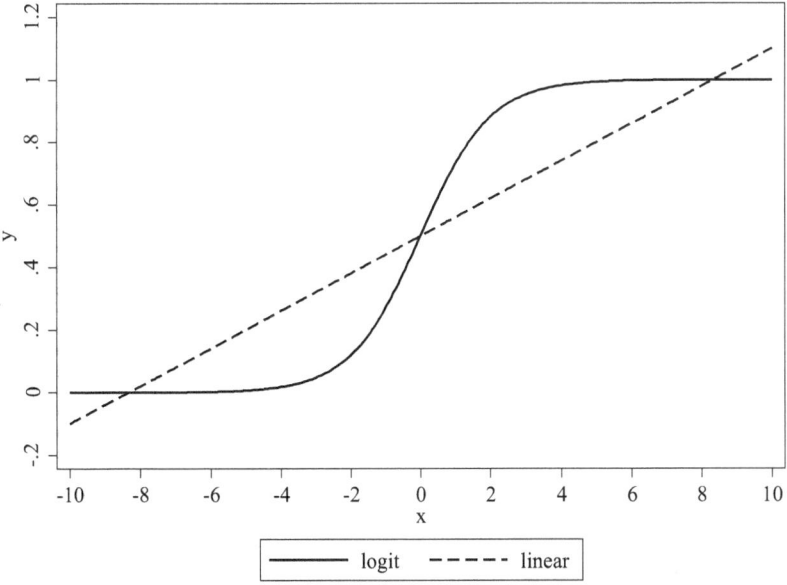

Fig. 8.2 Linear probability model and logistic regression model in comparison. (Own representation)

logit function, on the other hand, does not run linearly, but *s*-shaped. In the middle range of x-values (about between −3 and 3), the probabilities increase relatively strongly. The more the logit function approaches the extreme values of one or zero, the more the slope flattens out. It is said that the logit function asymptotically approaches the limit values, in this case 0 and 1, without ever crossing these boundaries.

This nonlinear course can also—and especially—be justified in terms of content. For example, let us assume that our dependent variable (y) measures whether a person owns a home or not and we want to predict this using the income of this person (x). In this case, it is not implausible that a certain minimum income is necessary to even think about buying a property. Starting with this threshold value, the probability of owning a home increases significantly. This increase then flattens out again in higher income ranges, because it is not so crucial for the home purchase whether a high or a very high income is available. Many relationships in empirical analyses follow a similar functional logic and therefore justify the use of the logistic instead of a linear regression model; this is probably the most important argument for using the logistic model.

8.1   Two Basic Concepts of Logistic Regression: Odds and Probability

To illustrate the concepts of probability and odds, we again use ALLBUS/GGSS data and first examine the bivariate relationship between the dependent variable non-denominationalism (0 = no, 1 = yes) and the current place of residence in East versus West Germany. Table 8.1 shows a corresponding cross table.

Table 8.1 Relationship between non-denominationalism and place of residence (observed frequencies with row percentages in parentheses). (Source: GGSS 2018)

	With denomination	Non-denominational	Total
West Germany	1782	598	2380
	(74.9)	(25.1)	(100.0)
East Germany	297	793	1090
	(27.3)	(72.8)	(100.0)
Total	2079	1391	3470
	(59.9)	(40.1)	(100.0)

Now, the *probability* can be calculated that people in Germany as a whole are without religious affiliation. To do this, one divides the number of people without religious affiliation (1391 people) by the total number of people in the sample (3470) and obtains a probability of 1391/3470 = 40.1% (see the row percentage value). In the next step, the probability of being without religious affiliation can now be determined separately for East and West Germany. In East Germany, this is significantly higher at 72.8% than in West Germany (25.1%).

In addition to the probabilities, alternatively the *odds* of being without religious affiliation can be determined. This odds is defined as the *ratio of two probabilities:* First, the probability that the dependent variable y takes the value 1 (a person is therefore without religious affiliation) and second, the complementary probability that the dependent variable takes the value 0, a person is therefore not without religious affiliation. Generally, this can be expressed as follows:

$$\text{odds} = \frac{P(y = 1)}{1 - P(y = 1)}$$

How are these chances calculated in Table 8.1? Based on all persons, the chance of having no denomination is the number of non-denominational persons (1391), divided by the number of persons with a denomination (2079), i.e. 1391/2079 = 0.67. The probability of being non-denominational (40.1%) is accordingly only about 0.67 times as large as the opposite probability of having a denomination (59.9%). One also speaks of a chance of "0.67 to 1".

Equivalent to the procedure for the probabilities, we can now determine this chance separately for East and West Germany: The chance of being non-denominational is 793/297 = 2.67 in East Germany and 598/1782 = 0.34 in West Germany. We note that there are considerably more persons in East Germany who are not affiliated with a denomination. The probabilityof having no denomination is about 2.7 times as high here as the probabilityof belonging to a church. In West Germany, on the other hand, this ratio is about 0.34 to 1.

In the context of logistic regression, a measure is now used that expresses the ratio of chances that were calculated separately for different groups or generally levels of an independent variable. This is the so-called *odds ratio*. As the name (odds ratio) suggests, two chances are set in relation here. For example, the odds ratio of being non-denominational for East versus West Germany is 2.67/0.34 = 7.9. The chance of being non-denominational is thus 7.9 times higher in East Germany than in West Germany. We can also 'turn around' the odds ratio and compare the West chance with the East chance. Here we get a result of 0.34/2.67 = 0.13. The chance of being non-denominational is thus only about 0.13 times as high in West Germany as in East Germany.

8.2   How to Interpret the Output of Logistic Regression?

So far, we have clarified by means of a simple example what is meant by a chance or probability in connection with a dichotomous dependent variable. Now the question arises, where these concepts are found within the logistic regression model. Before we clarify this, let us recall the linear probability model. Here, a linear regression is calculated on y; the regression weights indicate accordingly how the *probability* for y = 1 changes when the independent variable increases by one unit. However, we have already noted that the linear probability model has various problems with nominally scaled dependent variables. What is the procedure in logistic regression? The difference essentially consists in the fact that the primary dependent variable of logistic regression is not the probability for y = 1, but the *logarithmized chance*. The probability can also be determined by means of this method, but only after a computational intermediate step—more on this later.

Why is the *logarithmized* chance (the logit) used as the dependent variable in logistic regression and not, for example, the simple chance or the probability? To understand this, let us compare the value ranges of the three concepts discussed so far—probability, chance, logarithmized chance—based on Table 8.2 (cf. Kohler and Kreuter 2008, p. 265). Probabilities range between 0 and 1. Chances (odds) have a value range from 0 to infinity. Logarithmized chances can take values between minus infinity and plus infinity. For a probability of 0.5, the chance is exactly 1:1 and the logarithmized chance takes the value 0. For probabilities greater than 0.5, the chance is greater than 1 and the logarithmized chance is positive. If the probability is less than 0.5, the chance is less than 1 and the logarithmized chance takes negative values.

Table 8.2 Probabilities, odds and logits in comparison. (Own representation)

$P(y=1)$	Odds (odd) $=p(y=1)/1-p(y=1)$	Logit or ln(odd)
0.01	$1/99 = 0.01$	-4.60
0.05	$5/95 = 0.05$	-2.94
0.20	$20/80 = 0.25$	-1.39
0.50	$50/50 = 1.00$	0.00
0.80	$80/20 = 4.00$	1.39
0.95	$95/5 = 19.00$	2.94
0.99	$99/1 = 99.00$	4.60

We have already discussed above that a linear probability model can produce predicted values outside the permissible range [0,1]. For this reason, too,
the probability is ruled out as the primary dependent variable. The simple odds
are also only conditionally suitable for two reasons: First, it can also happen in
a regression on the odds that the predicted values fall below the lower limit (0).
Second, the odds are not symmetrical, as they change much more slowly in the
range below 1 due to the lower limit at 0 than in the open range above 1. In this
regard, probabilities, odds and logits are compared in Table 8.2.

The logit (the logarithmized odds) is very well suited for a regression model,
as it is not limited upwards or downwards and also symmetrical (around zero). For
this reason, the logit is used as the dependent variable in logistic regression:

$$\ln \left[\frac{P(y=1)}{1 - P(y=1)} \right] = \mathrm{Logit} = b_0 + b_1 x_1 \ldots + b_j x_j$$

Of course, the goal here is also to estimate the influence of covariates of interest.

The right side of this expression is familiar to us from linear regression, as it
is also a linear combination of a constant (b_0) and j regression weights (b_j) for the
independent variables x_j. One can therefore simplistically say that logistic regression is something similar to a linear regression on the logarithmized odds for
$y = 1$.

Using the logit, the probability for $y = 1$ can now be determined in a second
step. For this, it is only necessary to insert the logit (L) into the following formula, where "e" stands for the so-called Euler's number, which takes approximately the value 2.718:

$$P(y=1) = \frac{1}{1 + e^{-L}}$$

Let us return to the example to illustrate what we have just discussed. We are
still interested in how strongly the dependent variable non-religiosity (0 = no,
1 = yes) is associated with the place of residence in East Germany (coded as 1) or
West Germany (0). In Table 8.3 the corresponding output of a logistic regression

Table 8.3 Output of a bivariate logistic regression with the dependent variable non-religiosity (0 = no, 1 = yes). (Source: GGSS 2018, $n = 3470$)

	b	SE	z	p	e^b
East Germany	2.07	0.08	25.0	0.000	7.96
Constant	−1.09				0.34

with the most important elements is shown (details on the estimation of the coefficients follow below).

The term "*b*" denotes the logit coefficient, which is equivalent to the regression weight in linear regression. In general, the *b*-coefficient indicates how the logit of the odds for y $=1$ (here: non-denominational) changes when the independent variable increases by one unit. The logit coefficient of $b=2.07$ means accordingly that the logit of the odds of being non-denominational is 2.07 units higher in East Germany (x $=1$) than in the reference category West Germany (x $=0$).

Earlier, we had calculated various values by hand based on the cross table (Table 8.1): the probability and the odds of being non-denominational (separately for East and West) as well as the ratio of the East odds and the West odds of being non-denominational (odds ratio). We can now reproduce these indicators based on the output shown in Table 8.3.

Let us start with the odds of being non-denominational in East and West. To calculate these, we first need to determine the corresponding predicted values of the regression model (logit values). This is done, as in linear regression, by multiplying the constant and the regression weight. The logit value for East Germany is therefore $-1.09 + 2.07 = 0.98$. The constant refers, as in linear regression, to the case where the independent variable is equal to 0. The logit value for West Germany is therefore -1.09. Now we can exponentiate these logit values (e^b) and obtain the respective odds of being non-denominational (for East Germany $e^{0.98} = 2.67$, for West Germany $e^{-1.09} = 0.34$). If you now flip back a few pages, you will find that these are exactly the odds that we had calculated by hand based on the cross table (Table 8.1).

The odds ratio for the East-West comparison, which we had also calculated earlier, is shown in the output of the logistic regression in the column on the far right ($e^b = 7.96$). This value results from exponentiating the respective *b*-coefficient. For West compared to East Germany, the odds ratio is therefore $e^{2.07} = 7.96$. The odds of being non-denominational are therefore about 8 times as high in East Germany as in West Germany.

Finally, we need to clarify how the probability is calculated that East or West German people are non-denominational. To do this, we insert the logit value for East Germany (0.98) into the formula for calculating probabilities given above: $1/1 + e^{-0.98} = 0.728$. For West Germany, we calculate accordingly $1/1 + e^{1.09} = 0.251$. The probability of being non-denominational is therefore 72.8% (East) or 25.1% (West). Here, too, we arrive at the values that we had calculated by hand earlier.

What is the meaning of the other elements in the output in Table 8.3? When interpreting the standard error, it must be borne in mind that most empirical analyses (including ours) are based on sample data. The problem here is that statistical indicators that are calculated on the basis of samples usually vary around the 'true' value in the population due to random fluctuations (see Chap. 5). This also applies to our regression weight b. The standard error can be used to assess the reliability of estimates based on sample data. In general, three factors contribute to small standard errors and thus reliable estimates: a large sample, a strong effect of the independent variable on the dependent variable, and a large variance of the independent variable.

The standard error, however, is rarely interpreted by itself, but rather in combination with the b-coefficient to calculate the z-statistic (in STATA) or the Wald-statistic (in SPSS). The z-statistic, the equivalent to the t-statistic in linear regression, is simply calculated as the quotient of b-coefficient and standard error. In the Wald-statistic, which is χ^2-distributed, this quotient is additionally squared:

$$z = \frac{b}{SE_b}$$

$$Wald = \left(\frac{b}{SE_b} \right)^2$$

With the help of the z- or Wald-statistic, the null hypothesis is tested, similar to the t-statistic in linear regression, that the respective regression coefficient b in the population is equal to 0 (cf. Chap. 6). In the column "p" (or "Significance" in the SPSS output), the empirical error probability is reported, which exists when rejecting this null hypothesis. In our case, the East-West difference with regard to the probability of being without confession is over-random or significant. The probability of falsely rejecting the null hypothesis is less than 0.01%.

8.3 Probabilities, Odds, Log Odds and Average Marginal Effects: Guidelines for Result Interpretation

The logistic regression seems somewhat unclear, as depending on taste logit coefficients (b), odds ratios (e^b), predicted probabilities and, as we will see, further indicators can be used for result interpretation. It is necessary to pay close attention to the language, what can be inferred from the respective coefficients and what not. In the following section, different interpretation possibilities are summarized. In addition, a new concept, "average marginal effect", is introduced.

In Table 8.4 an overview is shown, which conclusions can be drawn from the b-coefficient and the odds ratio with regard to the odds, the log odds and the probability that the dependent variable takes the value 1. If the b-coefficient is positive, that is greater than 0, the log odds increase *exactly by b*, when the independent variable increases by one unit or decrease exactly by b, if the b-coefficient is negative. In the case of a positive b-coefficient, the odds ratio (e^b) takes values greater than 1. With a positive odds ratio, the odds for y $=1$ increase *by* the factor $|1 - e^b|$ (*to* e^b), when the independent variable increases by one unit or decrease by this value with an odds ratio less than one. To the probability for y $=1$, only can be said that it increases with a positive b-coefficient or an odds ratio greater than one. If the b-coefficient is negative and the odds ratio is less than one, the probability decreases accordingly. It is important, however, that it does *not* increase or decrease by b or by e^b!

To understand why this is the case, let us take a look at Fig. 8.3. Here again, the probability for y $=1$ is shown on the y-axis and a metric independent variable on the x-axis. In addition, two logit functions are drawn into the diagram. In the case of the solid line, the b-coefficient is positive ($b=0.5$) and in the case of the dashed line, negative ($b=-0.5$). The constant (b_0) is in both cases equal to 0 for simplification reasons.

It can be seen that the regression curve in the case of a negative b-coefficient falls from top left to bottom right and in the case of a positive b rises from bottom left to top right. It is important to note that these regression curves correspond to the predicted *probabilities*. Now it becomes clear why the probability does not increase exactly by b or decrease by b. This is due to the nonlinear shape of the logit function. For very small and very large x-values, for example, the solid curve rises only slightly, whereas in the middle range of x-values (about between -4 and 4), it rises sharply. Since the slope is not proportional over the range of x-values, it cannot be concluded from the b-coefficients (and also the odds ratios) by how much the probability for y $=1$ *exactly* increases when the

Table 8.4 Effects of positive and negative regression coefficients on the probability for $y=1$. (Own representation)

b	Odds-Ratio (e^b)	Logit/ln(odd)	Chance (odd)	Probability
$b>0$	$e^b>1$	Increases by b	Increases by $|1 - e^b|$	Increases
$b<0$	$e^b<1$	Decreases by b	Decreases by $|1 - e^b|$	Decreases

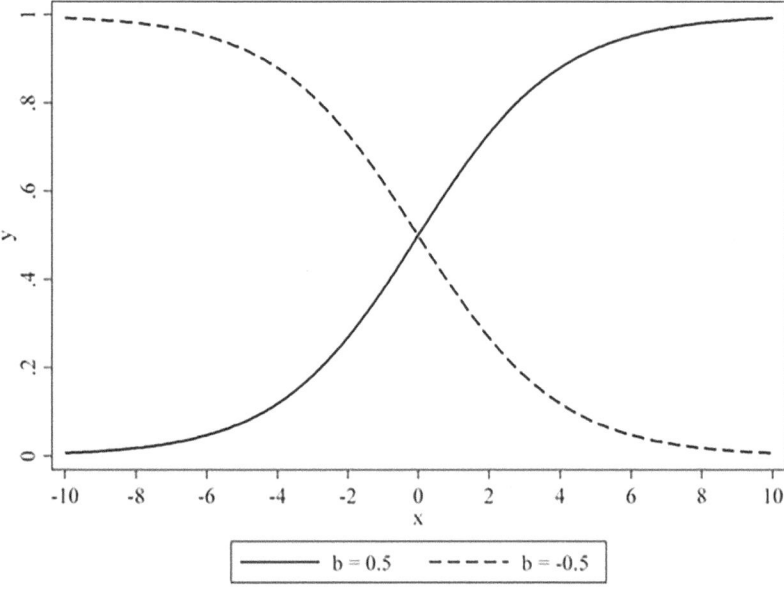

Fig. 8.3   Effects of positive and negative *b*-coefficients on the shape of the logit function. (Own illustration)

independent variable x increases by one unit. However, it is not wrong to state more generally that the probability increases with a positive *b* or decreases with a negative *b*. In the case of *b* = 0, the shape of the logit function corresponds, as in linear regression, to a horizontal line parallel to the x-axis.

Since logit coefficients and odds ratios do not allow exact statements about the change in the probability for y = 1, the concept of so-called Average Marginal Effects (AME; e.g. Hanmer and Kalkan 2013) is increasingly becoming established.[1] A Marginal Effect can first be generally defined as follows: How does the expected value of the dependent variable, i.e. here the probability for y = 1, change when the independent variable x increases by one unit and other independent variables are held constant (ceteris paribus)? In linear regression, these

---

[1]AME can be generated in Stata from version 12 using the command *margins, dydx(varlist)* (see *help margins*).

Marginal Effects simply correspond to the regression weights. As a result of the non-linearity of the logit function (Fig. 8.3), however, the Marginal Effects in logistic regression are not unique or constant. That is, they differ from unit to unit and thus depend on the specific values of the other independent variables that are in a model. The idea here is to proceed in a two-step process: First, the Marginal Effect for each unit with its specific values of the independent variables is calculated separately. Then, the mean of these different Marginal Effects is determined. This is then the Average Marginal Effect. It states that the probability for y = 1— on average of all observations of the present sample—increases by AME percentage points when the independent variable x increases by one unit (or marginally). AME thus give, in a nutshell, the average effect of x on the probability.

To summarize: In linear regression, the regression weight b indicates by how much the dependent variable changes exactly when the independent variable increases by one unit. It is also possible in logistic regression to make such exact statements, if one refers to log odds. The b-coefficients indicate how the log odds for y = 1 change exactly. With regard to the probability for y = 1, based on b or e^b, only generally can be stated whether this increases or decreases, without being able to give the exact value. To be able to make more precise statements also on the probability level, one can additionally use so-called Average Marginal Effects.

Excursus: How is the model fit determined in logistic regressions? In linear regression, the parameters of the regression equation, that is, the constant and the regression weight, are determined using the method of least squares (see Chap. 6). In logistic regression, on the other hand, it is a maximum likelihood (ML) estimation. This follows, simplified, the following logic: Given are the data of a sample, that is, for each person the information whether the dependent variable takes the value 0 or 1 and additionally the values of the independent variables. In the ML estimation, the regression coefficients are sought that make the occurrence of the sample data most likely. That is, regression coefficients should minimize a log-likelihood function and thus enable the best separation between the values of the dependent variable (1 and 0). In logistic regression, this is the log-likelihood function (LL), which is defined as follows:

$$LL = \ln(L) = \sum_{y_i=1} \ln(P(y_i = 1)) + \sum_{y_i=0} \ln(1 - P(y_i = 1))$$

The two sums that are added in this function correspond to the two values of the dependent variable (1 and 0). The function can take values between 0 (best possible model) and minus infinity. The value 0 is reached if, first, for all persons or cases where the dependent variable has the value 1, a probability of 1.0 is predicted. In this case, "$P(y=1)$" is always 1.0 and the logarithm of 1 is 0; the first sum is therefore equal to 0. For the entire LL function to take the value 0, the regression model must secondly predict a probability of 0 for all cases for which y $=0$ is observed. In this case, the expression "$(1-(P(y=1)))$" in the second sum is also 0, since nothing is subtracted from 1 and the logarithm of 1 again gives 0. In total, both sums are equal to 0 in this case and thus also the entire LL function.

The LL function, on the other hand, takes values smaller than 0 as soon as there are deviations between the observed y values and the predicted probabilities for y $=1$. For example, if y $=1$ is observed for a person and the regression model predicts a probability of 0.8, the residual, the error, is the value $1-0.8=0.2$. If, on the other hand, y $=0$ is observed for a predicted probability of 0.8, the residual is $0-0.8=-0.8$.

To better understand the calculation of the LL function, an example (with fictitious data) is shown in Table 8.5. The data set contains information for 15 persons

Table 8.5 Example for calculating the log-likelihood function value. (Own representation)

y	x	$P(y=1)$	$\ln(1-P(y=1))$	$\ln(P(y=1))$
0	2	0.23	−0.26	
0	4	0.46	−0.62	
0	1	0.15	−0.16	
0	5	0.59	−0.90	
0	7	0.81	−1.65	
0	2	0.23	−0.26	
0	1	0.15	−0.16	
1	4	0.46		−0.77
1	5	0.59		−0.52
1	7	0.81		−0.21
1	8	0.88		−0.13
1	5	0.59		−0.52
1	8	0.88		−0.13
1	9	0.92		−0.08
1	2	0.23		−1.47
			$\sum = -4.03$	$\sum = -3.83$

on the dependent variable y (with the values 0 and 1) and on a metric variable x. In the third column from the right, the probabilities for y = 1 predicted by the regression model are shown.

In the fourth and fifth columns, the values that enter into the two sums of the LL function are shown. For the person in the first row, for example, the value 0 is observed for the dependent variable. The logistic regression model, which includes the independent variable x, however, predicts a probability of 0.23 (so we make an error of −0.23). According to the formula for the LL function, we have to calculate "ln(1 − P(y = 1))" in this case, i.e. ln(1 − 0.23). The logarithm of 0.77 is −0.26, this value is entered in the fourth column from the left. Let us look at the person in the penultimate row of the table, who has the value 1 for the dependent variable. The predicted probability in this case is 0.92, so we make a relatively small error of 1 − 0.92 = 0.08. According to the formula for the LL function, the calculation here is "ln(P(y = 1))", i.e. ln(0.92) = −0.08. If we look at the other cases in the table, it becomes clear: The larger the deviations between observed values and predicted probabilities are, the larger negative values enter into the LL function. With the help of the two sums, which are given at the bottom of the table, we can now calculate the LL function value:

$$LL = \sum_{y_i=1} ln(P(y_i = 1)) + \sum_{y_i=0} ln(1 - P(y_i = 1)) = -4.03 - 3.83 = -7.86$$

Anyone who wants to replicate this result, enter the data from Table 8.5 into a data set and calculate a logistic regression. In Stata, the number −7.86 is displayed above the coefficient block under "Log likelihood", i.e. exactly the value we just calculated by hand. In SPSS, you can find in the output "Model Summary" a "−2 Log-Likelihood" value of 15.7 (calculation: −2 x −7.86 = 15.7). In the interpretation of the likelihood values in SPSS and Stata, there are the following similarities and differences: In both programs, the (theoretical) value 0 means that it is a perfect model without prediction errors, i.e. the prediction and observation values are identical. The *smaller* the LL value in Stata is, the larger are the deviations between prediction and observation values. In SPSS, on the other hand, the −2LL value becomes *larger,* the more or the larger residuals occur.

The LL values or −2LL values, which have a similar meaning as the sum of squares residuals in linear regression, are however difficult to interpret by themselves, as these values depend, for example, directly on the case number of the regression model and can therefore only be compared between different regression models with identical case number. The actual purpose of these function values is to use them to calculate various measures for assessing the model fit, the so-called pseudo-r^2 values. Simple in calculation and intuitively understandable is especially the pseudo-r^2 according to McFadden:

$$\text{Pseudo} - r^2(\text{McFadden}) = 1 - \frac{-2\text{LL}(\text{final model})}{-2\text{LL}(\text{intercept only model})}$$

For our regression model with the data from Table 8.5, we get:

$$\text{Pseudo} - r^2(\text{McFadden}) = 1 - \frac{15.714}{20.728} = 0.242$$

The pseudo-r^2 according to McFadden is a PRE ("proportional reduction of error") measure. The minimum value 0 means that the independent variables cannot improve the explanatory power of the model. The maximum value 1 is reached when the -2LL value in the model with independent variables (final model) is equal to 0 and accordingly no prediction errors would be made. Due to the PRE property, the coefficient indicates by how much percent the prediction of the dependent variable can be improved by the independent variables. This prediction improvement is 24.2% in our model.

In contrast to linear regression, there are different pseudo-r^2 variants for logistic regression, which are controversially discussed. A current overview can be found in Hemmert et al. (2018). Regarding the pseudo-r^2 according to McFadden, it should be said that this coefficient represents a rather conservative estimate for the model fit, i.e. it tends to take rather small values, but it has the advantageous PRE property.

In addition to the r^2 values, another indicator for the quality of the overall model is the likelihood-ratio-χ^2 (in SPSS in the output "Omnibus test" of the model coefficients under chi-square and in Stata on the right above the coefficient block under "LR chi2"). This is the equivalent to the F-test of linear regression. This test is based on the following null hypothesis: All coefficients of the regression model except the constant are equal to 0 in the population. The LR-χ^2 is always to be interpreted in relation to the degrees of freedom. These correspond to the number of regression weights in the model and are given in SPSS in the column "df" or in Stata behind "LR chi2" in brackets. From the LR-χ^2 value and the degrees of freedom, a significance level results, which can be interpreted as the probability of falsely rejecting the null hypothesis, i.e. that all regression coefficients in the population are actually equal to 0. The results show that this null hypothesis can be rejected with high probability. However, as with the F-test in linear regression (see Chap. 6), this is a rather weak test, which only refers to the overall model and says little about the explanatory power of the individual covariates.

8.4   An Example: Which Characteristics Influence the Probability of being Non-Denominational?

After all the technique, we now want to demonstrate the interpretation of the model results of logistic regression on a somewhat more detailed example. The dependent variable is again the dichotomous measurement of non-denomination-alism. The independent variables are gender, age, education, place of residence and marital status. At this point, the strength of logistic regression becomes clear. To test bivariate relationships (for example, between non-denominationalism and east/west), the tools that are available to us in the context of cross-tabulation analysis are usually sufficient (for example, the measures discussed above in Chap. 4, such as χ^2 or Cramer's V). The example shown above (Table 8.3) was limited to one covariate for didactic reasons. However, as soon as the effect of *several* independent variables on a nominal dependent variable is to be determined, the use of logistic regression is useful.

In Table 8.6 the results of a corresponding analysis can be found, which will now be discussed in detail. In the case of gender (coded as $0 =$ man, $1 =$ woman), the b-coefficient (-0.181) is negative. The logarithmic odds of being non-denominational are thus 0.181 lower for women than for men. This statement leaves

Table 8.6 Logistic regression model for the dependent variable non-denominationalism ($0 =$ no, $1 =$ yes; logit coefficients including standard errors, z values, odds ratios and average marginal effects). (Source: GGSS 2018, $n = 3434$)

	b	SE	z	e^b	AME
Woman	-0.181*	0.079	-2.3	0.83	-0.034
Place of residence: East Germany	2.115**	0.085	25.0	8.29	0.397
Years of education	0.085**	0.016	5.5	1.09	0.016
Marital status					
Divorced	0.163	0.137	1.2	1.18	0.031
Widowed	-0.489**	0.178	-2.8	0.61	-0.092
Married	-0.276**	0.093	-3.0	0.76	-0.052
Single (reference)	$-$	$-$	$-$	$-$	
LL (null model)	-2315.7				
LL (final model)	-1930.7				
Pseudo R^2 (McFadden)	0.166				

$\#p \leq 0.10$; $*p \leq 0.05$; $**p \leq 0.01$

us somewhat puzzled, because 'logarithmic odds' is an abstract technical term. Alternatively, the odds ratio can be interpreted: The *odds* of being non-denominational are only 0.83 times as high for women as for men, or 17% lower for women than for men. The average marginal effect (-0.034) allows a concrete statement at the level of the *probability*. The probability of being non-denominational, holding all other characteristics constant, is on average 3.4 percentage points lower for women than for men. The term "on average" is deliberately chosen here, as the marginal effects in specific subgroups, e.g. among older or younger, highly or lowly educated respondents, may be larger or smaller than 3.4 percentage points.[2] It is therefore an average marginal effect.

The difference between men and women is also significant at the 5% level; accordingly, we can reject the null hypothesis, which states that there is no gender difference in the population, with a high probability. The exact significance is reported in the respective statistical programs.

East Germans are more often non-denominational than West Germans. More precisely, they have an 8.29 times higher chance of being non-denominational than West German individuals.[3] On average, the probability of being non-denominational among East German respondents—ceteris paribus—is 39.7 percentage points higher than among West German respondents.

While gender and place of residence are dichotomous independent variables, educational level is metric. The chance of being non-denominational increases by about 0.09 times or 9% per year of education. The probability of being non-denominational increases on average by 1.6 percentage points per year of education. Comparing, for example, two individuals with 10 and 18 years of education, the probability of being non-denominational is on average 1.6 x 8 = 12.8 percentage points lower for the individual with 18 years of education than for the individual with 10 years of education.

Here one can easily make the mistake of considering the influence of education to be relatively weak, since b-coefficient and AME are relatively small and also e^b is only slightly larger than one. This interpretation would be wrong,

---

[2] Marginal effects in specific subgroups or for specific constellations can be generated in Stata within the *margins* command with the *at* option.

[3] When odds ratios greater than one are interpreted in percentages, errors often occur. The value 8.29 for the comparison between East and West Germany does not mean, for example, that the chance of being non-denominational in East Germany is 829% higher, but 729%. Since percentages greater than 100% intuitively make little sense, it is advisable here to speak of the "8.3-fold" or "8.3 times as high".

however, as these are exclusively unstandardized coefficients. Logit coefficient, odds ratio and AME each indicate how the logarithmic odds, odds or probability change *per increase of the independent variable by one unit.* While, for example, only one increase is possible for gender (from 0 = man to 1 = woman), educational level can increase relatively often (in the value range 8–18 years of education); this is why b-coefficient, e^b and AME are small in absolute terms. These are therefore unstandardized effects. Which explanatory factor has the strongest influence on the probability of being non-denominational cannot be reliably assessed on the basis of these coefficients.[4]

Let us return to the results: Marital status is a final example of a categorical variable. Compared to the reference category single, widowed respondents have a 39% lower chance of being non-denominational ($e^b = 0.61$). For married respondents, this chance is 24% lower than for singles ($e^b = 0.76$). The probability of being non-denominational is 9.2 percentage points lower for widowed respondents and 5.2 percentage points lower for married respondents than for singles. The difference between divorced and single respondents is not statistically significant.

The pseudo-r^2 is 0.17. It can be easily calculated from the log likelihood (LL) values at the end of the table: $1 - (-1930.7/-2315.7) = 0.166$. With regard to the pseudo-r^2, one should not have too high expectations—especially for the rather conservative variant according to McFadden. It is not very meaningful to specify minimum or target values, since r^2 strongly depends on the research question and in our case also on the number of covariates[5] included in the model. Pseudo-r^2 values close to 1 are hardly ever achieved in practice—and if they are, serious doubts about the model specification are usually justified. An improvement of the prediction of the dependent variable by almost 17% by four simple sociodemographic characteristics can therefore already indicate a good explanatory power of the model.

---

[4] Standardized effects can be generated in Stata for the logit model with the *spost13*-package by Scott Long and Jeremy Freese (see *help spost13*).

[5] Analogous to the corrected r squared in linear regression, there is the possibility to correct the McFadden-r squared by the number of model parameters (that is, the number of regression weights). To do this, simply subtract the number of model parameters from the log likelihood value of the final model. For our case, this results in: $1 - (-1936.7/-2315.7) = 0.164$. The value -1936.7 is obtained by subtracting 6 model parameters from the LL value of the final model.

One result, different presentations The output shown in Table 8.6 is very detailed, as it contains not only the b-coefficients but also the odds ratios (e^b), average marginal effects, standard errors and z-values. In scientific journals, more economical forms of presentation are usually found. Most often, the authors choose to present either b-coefficients, odds ratios or—increasingly—average marginal effects. Depending on the available space, taste or formatting guidelines in journals, standard errors, z- or Wald-values or only "asterisks" can be added, indicating the significance level. The reader should not be confused when studying the literature and always remember that the different possible forms of presentation are all based on the same procedure. However, even the term "logistic regression" is not always found in the literature, for example when a "logit model" or an "econometric model" is mentioned.

8.5   Pitfalls of Logistic Regression

Recently, some pitfalls have been pointed out that should be considered when applying logistic regression. Two aspects are closely related to the different forms of third variable influences that were discussed in Chap. 7: The change of the logit coefficients in stepwise (hierarchical) regression models (Mood 2010; Best and Wolf 2010, 2012) and the interpretation of interaction effects in logistic regression (Ai and Norton 2003).

In Chap. 7 it was explained in detail using the example of linear regression how hierarchical regression models can provide information about the relationship patterns between confounding variables (mediation or suppression). The following rule applied: If the effect of an independent variable X changes when introducing another third variable Z, this is a sure indication of a positive or negative correlation of X and Z. Unfortunately, this logic cannot be easily transferred to logistic regression models. Mood (2010) showed in a simulation study that the effect of an independent variable X can also change when introducing Z in logistic regression, even if X and Z are not correlated. According to Best and Wolf (2012), this causes distortions especially when the r^2 of the model is relatively large.

One first possibility to avoid this problem is to use the Average Marginal Effects. An important advantage of the AME is that the distortions described above occur much less frequently in hierarchical logistic regressions, i.e. only under extreme conditions such as very skewed variables (Best and Wolf 2012).

The second solution, proposed by Karlson et al. (2012), is as follows: The aim is to first calculate a reduced regression model of X on Y and then integrate the third variable Z into the model in a second step. To obtain unbiased results, one should additionally control for the *residuals of a regression of X on Z* in the reduced regression model (X on Y). These residuals can be generated as a new variable in the data set by specifying Z as the dependent variable, running a (linear or logistic) regression with the independent variable X on Z and using the option available in SPSS or Stata to save the residuals in the data set. This procedure is overall the most effective, as the distortions described above for calculating hierarchical regression models are completely corrected (Best and Wolf 2012).[6]

The second problem of the logistic regression model is related to the interpretation of interaction effects. If one refers to logits (logarithmized odds) or odds ratios, all rules for interaction effects that were discussed for linear regression in Chap. 7 also apply to the logistic model. With regard to the probabilities, which the hypotheses usually refer to, interactions, due to non-additivity and non-linearity in the logit model, can however lose their statistical significance and even change their sign.

To demonstrate this, the interaction effect of the variables place of residence (1 = East, 0 = West) and years of education (8 to 18, centered around the value 13 years of education) is shown in Table 8.7. The interpretation of the logit coefficients (*b*) and odds ratios (e^b) is done as in Chap. 7 explains: For example, the chance of being non-religious is 10 times higher for East Germans with 13 years of education than for West Germans. For West Germans, the chance of being non-religious increases by 12% per year of education. The interaction effect shows that the influence of education is about 10% lower for East Germans than for West Germans ($e^b = 0.9$). Both conditional main effects and the interaction effect are significant at least at the 5% level.

The results in Table 8.7 shows the results based on (logarithmic) odds. Can the interpretation now be directly transferred to probabilities? To test this, the Stata command *inteff* (Norton et al. 2004) is used. This calculates the Average Marginal Effect (AME) for the interaction effect[7] as well as the minimum and maximum of the interaction effect. Equivalent results are also obtained for the corresponding *z*-value. Part of the output is shown in Table 8.8. According to this, the average

---

[6] Another, even more elegant solution is to completely avoid hierarchical regression models and examine relationship patterns such as mediation or suppression via indirect effects in path models (see Chap. 7 and 9 for first hints).

[7] The command *margins, dydx(varlist)* mentioned above is not suitable for interaction effects. The corresponding cell for the AME was therefore deliberately left empty in Table 8.7.

Table 8.7 Output of a logistic regression with the dependent variable non-religiosity (0 = no, 1 = yes) and interaction effect. (Source: GGSS 2016 & 2018, $n = 6914$)

	b	z	e^b	AME
Place of residence East Germany	2.30**	35.8	10.00	0.408
Years of education (centered around 13)	0.11**	8.5	1.12	0.020
Years of education × East Germany	−0.10**	−4.0	0.90	
Constant	−1.28			

$+p \leq 0.10;\ *p \leq 0.05;\ **p \leq 0.01$

Table 8.8 Output of the Stata command "inteff" for the interaction effect "years of education × East Germany" from Table 8.7. (Source: GGSS 2016 & 2018)

	Mean	Minimum	Maximum
Average Marginal Effect	−0.0178	−0.022	−0.012
z-value	−3.26	−3.82	−1.67

interaction effect "years of education × East Germany" on the probability is −1.78 percentage points (with minimum −2.2 percentage points and maximum −1.2 percentage points). The z-value is on average highly significant −3.26 and ranges from −1.67 to −3.82.

The interaction effect can be interpreted—from the perspective of the education effect—as follows: According to the conditional main effect in Table 8.7, the probability of being non-religious increases *for West Germans* by an average of 2 percentage points (AME) with each year of education. According to Table 8.8, this increase in the probability of being non-religious per year of education is on average 1.78 percentage points weaker (AME for the interaction effect) for East German respondents.

The results in Table 8.8 appear surprising at first glance. How can an interaction effect have different values within the same sample and in a single regression model? This is, as already discussed, again a consequence of non-additivity and non-linearity in the logit model, i.e. the s-shaped course of the logistic regression curve on the probability level. In the present example, this means that the interaction "years of education x East Germany" is stronger and significant for certain cases in the sample and weaker and only marginally significant for other cases. Further information on this can be obtained from (not shown here) graphs that are automatically generated as part of the command *inteff*.

Overall, this results in the recommendation to examine interaction effects in logistic regression models in Stata in more depth with *inteff* and to calculate AME for interaction effects with this command.

8.6   Final Remarks

The importance of logistic regression in the social science research literature has been declining lately. The reason for this is certainly partly due to the "pitfalls" that were explained in Sect. 8.5. On the other hand, the Average Marginal Effects, which are enjoying increasing popularity, are very similar to the results of the linear probability model. Against this background, the procedure of calculating a logit model and then determining AME seems quite cumbersome. The linear probability model, which approximates the AMEs well, is comparatively simpler and is also experiencing a renaissance for this reason (Breen et al. 2018). Simulations by Best and Wolf (2012) show, however, that the linear probability model can lead to biased results with strongly skewed x-variables. Moreover, the argument that the non-linear course of the logit function is often more appropriate still applies. Against this background, logistic regression—applied adequately—will retain its place in the methodological toolbox.

Finally, some literature references: Compact introductions to logistic regression are provided by Pampel (2000), Menard (2001) and Borooah and Lewis-Beck (2001) as part of the Sage study script series. The standard work for regression models with categorical dependent variables in Stata is by Long and Freese (2014).

References

Ai, Chunrong, and Edward C. Norton. 2003. Interaction terms in logit and probit models. *Economics Letters* 80:123–129. https://doi.org/10.1016/s0165-1765(03)00032-6.

Best, Henning, and Christof Wolf. 2010. Logistische Regression. In *Handbuch der sozialwissenschaftlichen Datenanalyse*, Eds. Christof Wolf and Henning Best, 827–854. Wiesbaden: VS Verlag. https://doi.org/10.1007/978-3-531-92038-2_31.

Best, Henning, and Christof Wolf. 2012. Modellvergleich und Ergebnisinterpretation in Logit- und Probit-Regressionen. *Kölner Zeitschrift für Soziologie und Sozialpsychologie* 64:377–395. https://doi.org/10.1007/s11577-012-0167-4.

Borooah, Vani K., and Michael S. Lewis-Beck. 2001. *Logit and probit: Ordered and multinomial models. Quantitative applications in the social sciences 138*. Thousand Oaks: Sage. https://doi.org/10.4135/9781412984829.

Breen, Richard, Kristian Bernt Karlson, and Anders Holm. 2018. Interpreting and understanding logits, probits, and other nonlinear probability models. *Annual Review of Sociology* 44:39–54. https://doi.org/10.1146/annurev-soc-073117-041429.

Hanmer, Michael J., and Kerem Ozan Kalkan. 2013. Behind the curve: Clarifiying the best approach to calculating predicted probabilities and marginal effects from limited dependent variable models. *American Journal of Political Science* 57:263–277. https://doi.org/10.1111/j.1540-5907.2012.00602.x.

Hemmert, Giselmar A. J., Laura Marie Edinger-Schons, Jan Wieseke, and Heike Schimmelpfennig. 2018. Log-likelihood-based pseudo-r2 in logistic regression: Deriving sample-sensitive benchmarks. *Sociological Methods & Research* 47:507–531. https://doi.org/10.1177/0049124116638107.

Karlson, Kristian Bernt, Anders Holm, and Richard Breen. 2012. Comparing regression coefficients between same-sample nested models using logit and probit: A new method. *Sociological Methodology* 42:286–313. https://doi.org/10.1177/0081175012444861.

Koch, Achim. 1994. Teilnahmeverhalten beim ALLBUS 1994. Soziodemographische Determinanten von Erreichbarkeit, Befragungsfähigkeit und Kooperationsbereitschaft. *Kölner Zeitschrift für Soziologie und Sozialpsychologie* 49:89–122.

Kohler, Ulrich, and Frauke Kreuter. 2008. *Datenanalyse mit Stata: Allgemeine Konzepte der Datenanalyse und ihre praktische Anwendung.* München: Oldenbourg.

Konietzka, Dirk, and Michaela Kreyenfeld. 2005. Nichteheliche Mutterschaft und soziale Ungleichheit im familialistischen Wohlfahrtsstaat. *Kölner Zeitschrift für Soziologie und Sozialpsychologie* 57:32–61. https://doi.org/10.1007/s11577-005-0110-z.

Long, Scott, and Jeremy Freese. 2014. *Regression models for categorical dependent variables using Stata.* College Station: Stata Press.

Menard, Scott W. 2001. *Applied logistic regression analysis. Quantitative applications in the social sciences 106.* Thousand Oaks: Sage. https://doi.org/10.4135/9781412983433.

Mood, Carina. 2010. Logistic regression: Why we cannot do what we think we can do, and what we can do about it. *European Sociological Review* 26:67–82. https://doi.org/10.1093/esr/jcp006.

Norton, Edward C., Hua Wang, and Ai Chunrong. 2004. Computing interaction effects and standard errors in logit and probit models. *The Stata Journal* 4:154–167. https://doi.org/10.1177/1536867x0400400206.

Pampel, Fred C. 2000. *Logistic regression: A primer. Quantitative applications in the social sciences 132.* Thousand Oaks: Sage. https://doi.org/10.4135/9781412984805.

Szydlik, Marc, and Jürgen. Schupp. 2004. Wer erbt mehr? Erbschaften, Sozialstruktur und Alterssicherung. *Kölner Zeitschrift für Soziologie und Sozialpsychologie* 56:609–629. https://doi.org/10.1007/s11577-004-0106-0.

An Outlook on Advanced Statistical Analysis Methods

9

In the context of this book, two widely used statistical evaluation methods, linear and logistic regression, have already been discussed. This has an essential reason: Those who have understood the functioning of these methods usually also find faster access to advanced methods in the field of social science data analysis. Some of these complex methods are briefly and overview-like presented in the following chapter. The necessity of this undertaking becomes quickly apparent when one takes a look at leading sociological journals, in which elaborate statistical methods, such as multilevel analysis, are now very common. Anyone who wants to understand and possibly criticize the current research, therefore, increasingly depends on appropriate statistical expertise.

Two goals are pursued in the following chapter: First, the reader should be brought closer to the type of research question for which the respective method is needed or which evaluation possibilities are available in principle. The focus is on conceptual aspects of the respective method, not on the technical details. However, these two aspects cannot always be completely separated from each other, which makes a certain balancing act between too much and too little depth of detail unavoidable. Second, the reader is referred to introductory and advanced textbooks on the respective method.

Due to the multitude of statistical methods, a selection was made that is oriented to the frequency of using the respective methods in the current sociological research. The following methods are discussed—each using example questions—methods for 1) event data analysis, 2) multilevel analysis with cross-sectional data, 3) causal analysis with panel data, 4) path models and 5) meta-analyses.

9.1   Event Data Analysis

The subject of event data analysis are, as the name suggests, events such as marriage, divorce or death, which can be located in time. An event is defined as the transition from one discrete state, such as married or employed, to another, divorced or unemployed. A distinction is made between an absorbing target state, in which a further state transition is not possible, such as death, and recurring events, which can occur several times in the life course, such as unemployment. Event data provide information about the exact points in time at which events occur. The time interval between two adjacent discrete states, such as from married to divorced, is called an episode.

Another important term is the duration. It indicates how long an individual remains in the initial state, such as married, until an event, the divorce, occurs or the observation time ends. However, in order to experience an event, persons must be in an initial state that allows the transition to the corresponding target state. For example, a single person cannot become a widow or a non-employed person unemployed. The group of persons who are exposed to the risk of state change at a certain point in time is called the "risk set".

Now that central terms in event data analysis have been defined, we can use some examples. What questions can be investigated with the help of event data analyses? Table 9.1 shows an overview. In the family sociological area, for example, the transition to the first marriage can be analyzed. Persons can experience this event—they are "at risk"—when they reach the minimum marriage age of 18 years. The "risk set" accordingly consists of persons in the initial state single from 18. The time between the 18th year of life and the entry of the target state, the marriage, or the end of the observation time, corresponds to the duration.

Table 9.1   Example questions for event data analysis. (Own representation)

Life domain	Event	"Risk-Set"	Process start
Partnership and family	First marriage	Single persons	Minimum marriage age
	Family formation	Childless persons	About age 14
	Divorce	Married persons	Marriage
Employment	Unemployment	Employed persons	Start of employment
	End of parental leave	Women with birth	Birth of the child
Mortality	Death	All people	Birth

Depending on the accuracy of the data, this time can be in years or months. If a person marries exactly at the end of their 30th year of life, the duration in the initial state "single" is 12 years or 144 months.

To answer the question of what specific advantage event data analysis offers, the concept of censoring must first be introduced. One distinguishes between left and right censoring. In the case of left censoring, the time of the process start is unknown. As an example, think of an event data analysis of the transition of a couple to cohabitation. The process start could be defined in this case by the beginning of the partnership. If this date is not determinable for some reason, for example refusal to answer or disagreement of the partners, this is a case of left censoring. This form of censoring is rather rare in practice, but it brings with it great methodological problems, as it makes the calculation of event data analyses much more complicated.

The normal case in the context of event data analyses, on the other hand, is right censoring. This occurs whenever a person has not experienced the event of interest at the time of the survey, i.e. the corresponding episode is not yet completed. For example, in the analysis of divorces, most married couples are still intact at the time of data collection. Possibly some of these couples will divorce in the future; however, this is not known at the time of the survey. The duration of stay is right censored in this case.

Event history analysis makes it possible to handle right censoring statistically adequately. Here is an example: Suppose we drew a random sample of persons of marriageable age in 2021. The transition to the first marriage is to be examined. A first idea would be to analyze only those persons who were already married at the time of the survey. One possibility could be to calculate a linear regression on their age at marriage. However, such an approach is problematic insofar as the group we chose is very likely to be selective. Here, persons who have characteristics such as, for example, a divorce of their parents, which favor an early marriage, will be overrepresented. The results obtained in this analysis would probably be biased and misleading, as they are not generalizable to the entire group of persons of marriageable age. If, on the other hand, we calculate an event history analysis, we take into account not only the already married persons, but also those who are still single at the time of the survey. These right-censored cases, however, are by no means comparable to missing values. The fact of not being married at a certain age is an important piece of information, even if the event of interest has not yet occurred in this case. By appropriately handling right censoring, it is therefore also possible to analyze persons from young birth cohorts, who have a relatively high proportion of right-censored cases, but are particularly important for estimating future trend developments.

The dependent variable in event history analysis always consists of two pieces of information: First, it must be known whether the event of interest has already occurred or not for the respective person. Second, it is important how long a person stays in the initial state until the event occurs or the observation time ends due to right censoring. The second piece of information makes the crucial difference between simple cross-sectional and event history data.

On the basis of the event indicator and the duration, various function values are calculated in event history analysis. The two most important ones, the survival function and the transition rate, the so-called hazard rate, are now conceptually introduced using an example.

In Fig. 9.1, so-called transition rates are shown, which refer to the transition to family formation. The sample consists of 15–42 year old persons who are childless in the initial state. The lines correspond to the conditional probability that persons in the respective age group make the transition to the first child. The condition is that they are childless up to this age. A hazard rate of about 11 % for women in the age group 31–34 years means, for example, that out of 100 women, on average 11 give birth to a child in the next time unit (here: between the 31st and 34th year of life).

Using Fig. 9.1, two essential questions can be illustrated that are of central interest in the context of event data analysis. First, it is about the relation-

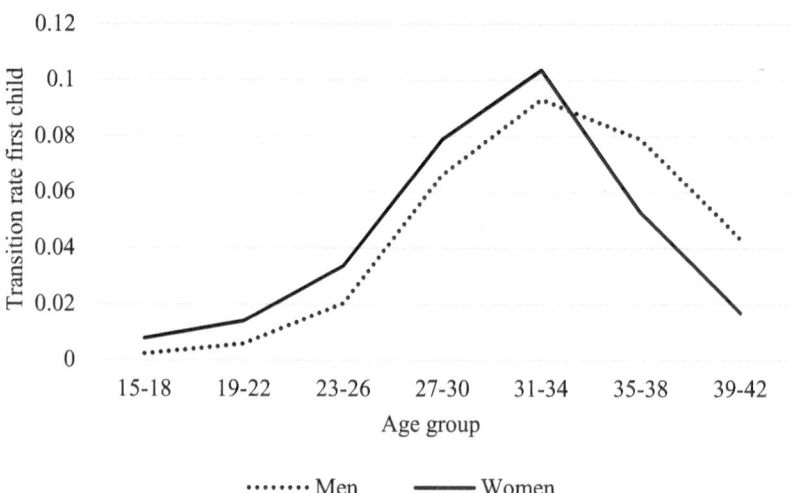

Fig. 9.1 Transition rates to the first child for men and women. (Source: German Family Panel, Release 10.0)

ship between the duration of stay (here: the age) and the probability of the event occurrence. The graph indicates a bell-shaped relationship. Around the 30th year of life, the hazard function takes the highest values. Younger and older people, on the other hand, have a lower conditional probability of making the transition to the first child. This so-called duration dependence, that is, the exact pattern of the hazard rate, is an important topic in event data analysis, as it also guides the choice of a suitable event-analytic model.

The bell-shaped pattern observed in Fig. 9.1 suggests that the conditions for starting a family are favorable towards the end of the third decade of life, while there seem to be factors in the age ranges before and after this period that inhibit the fertility process. For example, many young people delay starting a family because they have not yet completed their vocational training.

However, which characteristics exactly play a role here cannot be recognized solely on the basis of the pattern of the hazard rate. Here we come to the second central question in event data analysis: Are there individual factors such as place of residence, gender, education, employment status, which influence the probability of a birth positively or negatively? This question can be answered with the help of event-analytic regression models, in which the different covariates affect the transition rate.

The two lines drawn in Fig. 9.1 refer to two groups: male and female respondents. The effect of gender can now have two effects. On the one hand, it is conceivable that, for example, women have a higher probability of starting a family across all ages than men. The difference in the line courses would in this case be expressed in a shift on the vertical (y) axis. The 'bell' of the women would therefore be above that of the men across all ages and one would speak of a "level effect". On the other hand, it is conceivable that gender only influences the timing of family formation ("timing effect"). This could manifest itself, for example, in the fact that the transition rate for women increases earlier, that is, in lower age ranges, than for men. When looking at the courses in Fig. 9.1, it becomes clear that there is clearly a timing effect here. The bell related to women is shifted to the left on the x-axis, compared to men. Women therefore have a slightly higher probability of starting a family in lower age ranges. In the age range over 30, however, the women are overtaken by the men. Overall, it can thus be summarized that men tend to start a family somewhat later in the life course than women.

To what extent level or timing effects are present can be better assessed with the help of Fig. 9.2. Here, so-called survival functions are shown. They indicate in the present case the proportion of persons who are still childless at the respective age, i.e. who have not yet experienced the event under investigation at the respective point in time. For example, about 50 % of women are still childless at

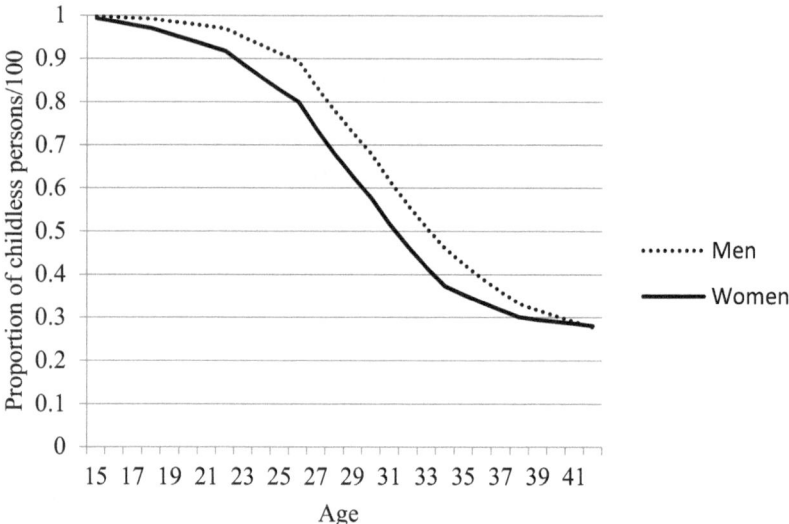

Fig. 9.2 Survival functions for family formation of men and women. (Source: German Family Panel, Release 10.0)

the age of 30. A comparison of the survival functions now clarifies that women initially make the transition to family formation faster in lower age ranges and men catch up later. At the age of 42, men and women are finally about equal with a share of childless persons of about 28 %. Since there is no gender-specific difference at the end of the observation window, this is a pure timing effect.

For the inexperienced user, the entry into event data analysis is usually complicated by the fact that there are a variety of different model classes. After the introduction of some essential terms, a brief overview of the different variants follows. A basic distinction concerns event data analyses for time-continuous event data and the so-called discrete-time event data analysis. Methods for time-continuous event data are used when a very precise, for example month-accurate[1],

---

[1]What is meant by an exact measurement of the duration of stay depends strongly on the research question. For example, if the time between the onset of a disease and the return to work is given in months, this is certainly a very inaccurate measurement. The month-accurate indication of the duration of marriage until divorce, on the other hand, seems sufficiently exact.

measurement of the duration of stay is required. Since the collection of time-continuous data can be costly and may encounter resistance in a concrete interview situation, the duration of stay is often measured imprecisely. This would be the case, for example, if only the calendar year of a divorce is known, but not the month of divorce. In the case of such grouped measured durations of stay, the discrete-time event data analysis is used. This method is also the standard method for the evaluation of panel data. In a panel, durations of stay are often measured in discrete or grouped form, since persons are repeatedly interviewed at regular intervals, for example one year.[2]

Furthermore, a distinction has to be made between nonparametric and parametric methods of event data analysis. In a parametric method, the user assumes a certain course of the transition rate by choosing a specific mathematical function, for example bell-shaped or monotonically increasing. In nonparametric methods, such as life tables or Kaplan-Meier estimators, this is not the case, as the observed course of the transition rate is simply explored. However, nonparametric methods quickly reach their limits when the simultaneous influence of several covariates on the transition rate is to be examined. For such complex regression models, parametric models have to be used. Here, it has to be checked to what extent the course of the transition rate assumed by the model fits the actually observed data. Some parametric methods, such as the so-called generalized log-logistic model (Brüderl and Diekmann 1995), have a real added value, as they allow an explicit differentiation between timing and level effects.

Finally, without claiming to be exhaustive, we will discuss relevant literature on event data analysis. Compact, introductory presentations on event analysis can be found, for example, in Allison (2014), Blossfeld et al. (2019) or.Mills (2011). The standard work for discrete-time event data analysis is Singer and Willett (2003, p. 357–463). Applications in the software Stata are described in detail by Blossfeld et al. (2019) and Cleves et al. (2010). For the software R, Broström (2012) provides a guide.

---

[2]Event data based on panel surveys are often left-truncated. This refers to constellations where persons were already in the risk state for a certain time before the start of the first panel survey. Time-discrete event data analyses also lead to an unbiased estimation of the transition rates in this data structure (see Guo 1993). Time-discrete event data analyses are therefore the standard method for event analyses with panel data.

9.2   Hierarchically Structured Data: Multilevel Analysis

In sociology, it is emphasized that social action does not only depend on individual characteristics such as gender, education or personality, but also on the social context in which the actor is embedded. When context information is available in social science data sets, these are hierarchically structured data. Fig. 9.3 illustrates this using an example. Here, students on level 1 are grouped into school classes or schools. Other examples of grouping variables on level 2 are countries or households.

Methods for analyzing hierarchically structured data sets are summarized under the umbrella term of multilevel analysis. To illustrate the functioning of these methods, we use a highly simplified example: survey respondents (level 1) group in countries (level 2). The dependent variable to be analyzed is religiosity, measured on a scale of 0–10 (10 = very religious).

The religiosity of the respondent can depend on influences at the individual level, for example religiosity of the parents, and at the country level, for example religious culture of a country. Characteristics at the country level are the same for all inhabitants of a country, but can differ between countries. The relationship between the religiosity of the parents and the religiosity of the respondent is now examined for a fictitious data set with 40 people from two countries with 20 respondents each. In Fig. 9.4 a scatter plot for all 40 respondents is shown.

The drawn regression line corresponds to the *cross-country* relationship between the religiosity of the parents and individual religiosity. Here, the fact that there are two different context units (two countries) is ignored, since only one regression analysis is calculated for the entire data set. For each unit that the scale "religiosity of the parents" increases, individual religiosity increases by 0.35 units

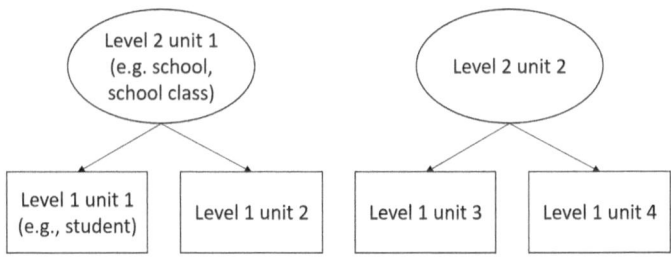

Fig. 9.3   Examples of hierarchically structured data. (Own representation)

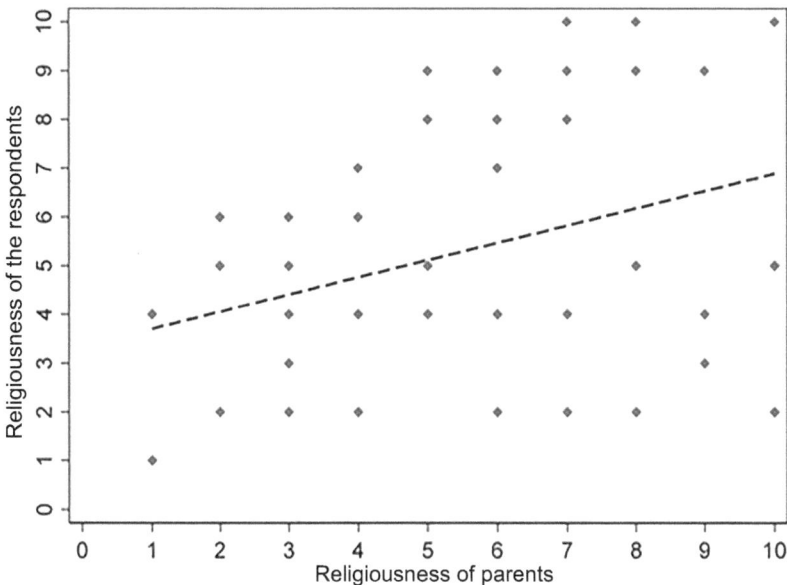

Fig. 9.4 Scatter plot of the relationship between religiosity of the parents and religiosity of the respondent. (Own representation)

(regression weight or slope). If the scale for religiosity of the parents takes the value zero, the mean religiosity of a respondent is 3.35 (intercept).

Since the regression analysis in Fig. 9.4 neglects the hierarchical structure of the data, that is, the nesting of the 40 people in two countries, it leaves two questions unanswered: First, are there possibly differences in the mean religiosity between the countries? Second, is the direction or strength of the relationship between the religiosity of the parents and the individual religiosity in one country different from the other?

To answer these questions, two separate regression analyses, one per country, are performed in a next step. The results, which are shown in Fig. 9.5, reveal two clear differences between the countries: First, it can be seen that the mean religiosity in the first country is lower than in the second. In the regression equation, this is expressed by the different intercepts (2.18 versus 3.73). The second difference is that the relationship between the religiosity of the parents and the individual religiosity in the second country is much stronger than in the first. This can be

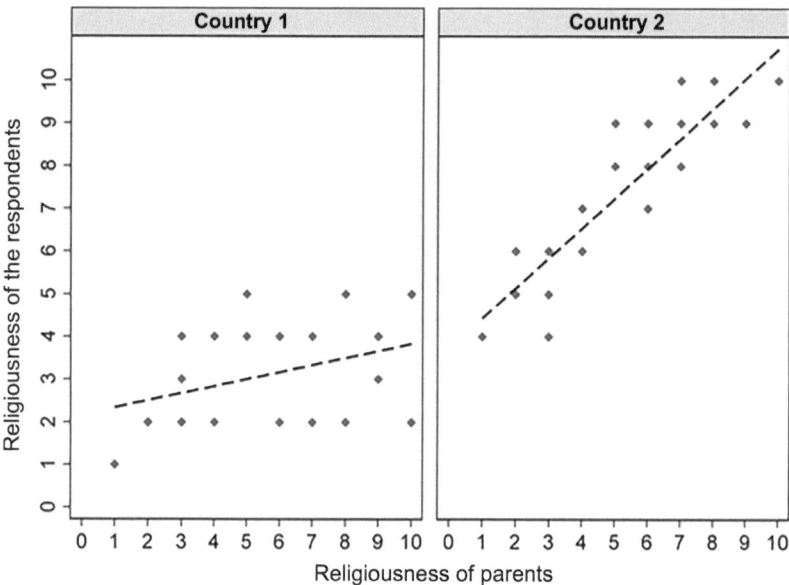

Fig. 9.5 Scatter plots and regression line of the relationship between the religiosity of the parents and individual religiosity in two countries. (Own representation)

read from the slopes (0.16 versus 0.70). The corresponding regression equations are thus $y = 2.18 + 0.16\ x$ in country 1 and $y = 3.73 + 0.70\ x$ in country 2.

By calculating two country-specific regression models, we have now performed a very simple form of multilevel analysis. The differences between the two context units on the second level, the two countries, can be identified by the different intercepts and slopes. If the data set now consists of very many context units, for example 30 countries, the procedure presented—calculation of a separate regression analysis for each context unit—becomes impractical. More complex methods of multilevel analysis solve this technically more elegantly, but follow the same principle. Here too, differences between the context units are determined by the varying intercepts and slopes.

So if, to return to the example, there are differences between the two countries in terms of the average religiosity (intercept) or the relationship between the religiosity of the parents and the religiosity of the respondent (slope), the question arises as to possible explanatory factors. It is conceivable that the country-specific differences in the regression constants—average level of religiosity—are due to individual characteristics of the respondents, for example their level of education. In addition to such explanatory factors at the individual level, however, country characteristics, such as a different degree of modernization, may also be responsible for the differences between the countries. Furthermore, the relationship between the religiosity of the parents and the individual religiosity in the second country could be stronger because the intergenerational relationships are closer here, in that parents and children see each other more often or feel more emotionally connected.

In multilevel analyses, it is therefore checked whether there are, in addition to the effect of individual factors, properties of the social context that are explanatory for an individual characteristic, here religiosity. The dependent variable, which is basically located at the individual level, is examined simultaneously at two levels of analysis, i.e. at the individual and context level. The advantage here is to relate the individual and context effects to each other. Does the effect of an individual characteristic change depending on the context properties?

So-called multilevel models with random coefficients, which now dominate the research, can now be divided into different variants. A first criterion of distinction concerns the measurement level of the dependent variable. Thus, multilevel analyses can be performed not only for metric but also for dichotomous dependent variables (Guo and Zhao 2000)—and thus also within the framework of event data analyses. Another criterion of distinction concerns the question of whether only level differences between the context units are examined, so-called "random intercept" models, or additionally interactions between individual and context characteristics and thus so-called "random slope" models.[3]

Some examples can make it clearer which questions are investigated in the current sociological research with multilevel analyses. Schulze et al. (2009) examine, among other things, the probability that students receive a recommendation for a grammar school. This is a three-level analysis, as students (level 1) are grouped in school classes (level 2) and school classes in schools (level 3). The probability of receiving a grammar school recommendation depends, on the one hand, on individual characteristics. For example, students with a good average grade and children of parents with a high social status receive a grammar school recommendation more often. In addition, there are effects of the social context,

---

[3] In addition, it must be taken into account that the inferential statistical significance tests are biased when conventional analysis methods such as linear regression are applied to hierarchically structured data. This problem is also avoided in the context of multilevel analyses with random coefficients.

here the respective class and school. The probability of a grammar school recommendation increases in addition to the individual effects with the social status level of the classes and schools.

Hank et al. (2004) examine the influence of regionally available childcare facilities on the fertility behavior of West and East German women. In this analysis, the first level corresponds to individuals and the second to districts. Level differences in the probability of birth between the districts are now attributed, in addition to various individual characteristics such as education or employment status, to how many institutionalized possibilities of childcare (crèches, kindergartens, after-school care) are available in the respective district. The result shows that the availability of childcare promotes the transition to the first birth in East Germany. The study mentioned is thus a combination of an event data and a multilevel analysis.

A compact and very understandable introduction to multilevel analysis is provided by Luke (2019). As a selection of further textbooks on the topic, Snijders and Boskers (2011), Hox (2010) and Goldstein (2010) are recommended. The standard work for implementing multilevel analyses in Stata, also with longitudinal data, is by Rabe-Hesketh and Skrondal (2022). Garson (2019) explains the practical application in five software programs (Stata, SPSS, SAS, R and HLM).

9.3   Causal Analysis with Panel Data

Panel data, which also have a hierarchical structure, are repeated measurements of one or more variables for the same individuals. Panel data have several advantages over cross-sectional data: They are more informative, as they allow the analysis of individual developments. This is made possible by the fact that, in addition to the differences between individuals, information on differences within individuals over time is also available. In doing so, information on the temporal—and possibly also causal—sequence of states and events is available. Particularly advantageous in this context is that panel data allow the control of unobserved individual heterogeneity, which is a major problem for the identification of causal effects with cross-sectional data.

Using the example shown in Fig. 9.6, it is illustrated which different types of effects can be differentiated with panel data. The basis is a fictitious data set of five individuals with partners who are single in the first panel wave. The dependent variable in this example is the frequency of church attendance, measured on a scale between 0 and 10 (10 = very frequent). The aim is to investigate whether marriage causally influences the subsequent frequency of church attendance.

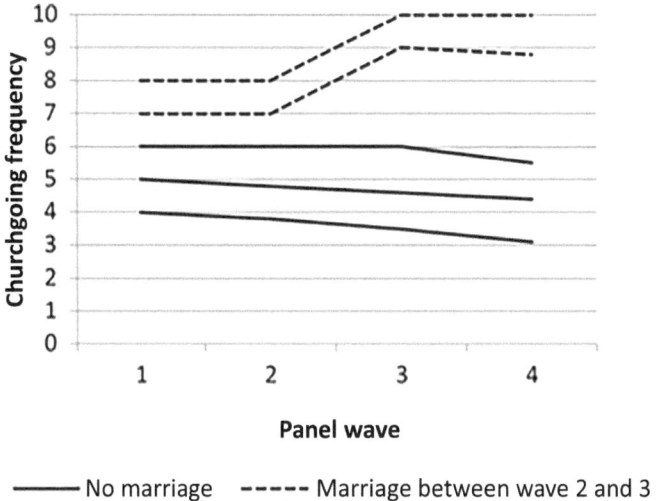

Panel wave

———— No marriage   - - - - - Marriage between wave 2 and 3

Fig. 9.6 Change in the frequency of church attendance of five fictitious individuals over four panel waves. (Own representation)

The figure shows the development of church attendance frequency of five persons over four panel waves. The dashed lines represent persons who marry between wave two and three. The solid lines represent persons who do not make the transition to marriage within the observation period. The figure suggests that three effects are present: First, there is a weak period effect, as the frequency of church attendance tends to decrease over time for all persons. Second, there is a positive (causal) effect of marriage on religious practice. Those persons who marry between the second and third wave go to church much more often after the marriage. In the control group without marriage, this effect is not observed. Third, there are indications of self-selection. Those persons who marry have on average a higher frequency of church attendance before the marriage than those who do not marry.

How can we proceed to determine the causal effect of marriage? Let us assume that we only had cross-sectional data, for example from the third wave, available. Here, one possibility would be to compare the mean frequency of church attendance of the two already married persons with the average value of the three single persons. It is obvious that in this case we would overestimate the causal effect of marriage. This is due to self-selection. Those persons who marry go to church more often before the marriage. The comparability of the "experimental" and control group is therefore not given. This is a typical problem of cross-sectional data.

What advantage do we have if we use the panel data, i.e. longitudinal information? Here, there is the possibility to compare those persons who marry with themselves and thus perform a so-called within-estimation. We consider the mean frequency of church attendance before the marriage in relation to the average value after the marriage. This is already a much better estimation of the causal marriage effect. However, we have already noted that the frequency of church attendance decreases for all persons over time. To determine the causal influence of marriage exactly, we have to take this period effect into account in the estimation.

This is possible within the framework of the so-called "difference in difference" method. As an example, we base our calculation on a comparison of the second and third panel wave:

$$\frac{(10-8)+(9-7)}{2} - \frac{(4.6-4.8)+(3.5-3.8)+(6-6)}{3} = 2-(-0.17) = 2.17$$

The right fraction in the displayed formula refers to the three persons who do not marry. Here, the value of the frequency of church attendance in the second wave is subtracted from the value in the third wave. The left fraction refers accordingly to the two persons who marry between the second and third wave. Here, the difference in the frequency of church attendance between the time before the marriage (second wave) and after the marriage (third wave) is formed. Subsequently, the difference of the difference between the control and "experimental group" is calculated. The causal marriage effect is thus 2.17.[4]

The small example illustrates that the main advantage of panel data is the generation of variance within persons. In this way, quasi-experiments can be conducted. How does the dependent variable change for persons who experience certain events between the panel waves ("quasi-experimental groups") or not (comparison group)? In contrast to cross-sectional data, it is possible within the framework of a panel to determine the causal effect of a "treatment" (for example, an event) on various dependent variables at least approximately.

In the recent literature on panel analysis (an overview can be found below), two (linear) regression methods for panel data are mainly discussed: the random-

---

[4] Even in the difference-in-difference estimation, it is not absolutely certain whether the effect of marriage is actually causal. There is the possibility that the two persons who marry have experienced unobserved changes between the second and third wave that simultaneously affect the probability of marriage and the frequency of church attendance. An example would be the birth of a child.

effects model (RE-model) and the fixed-effects model (FE-model). The RE-model[5] uses both sources of variance that are available in the context of panel data: the variance between persons and the variance within persons. However, this method has the decisive disadvantage that it is susceptible to self-selection effects, which makes the identification of causal effects much more difficult (Allison 1994; Halaby 2004). Therefore, the application of the FE-model is recommended. This method works conceptually similar to the difference-in-difference estimation presented above, uses only one source of variance (within persons) and is more robust against self-selection effects.

The application of the FE-model is now finally demonstrated using a small example (Table 9.2) (detailed results can be found in Lois 2011). The data are based on the Socio-Economic Panel for the period 1992–2007. For the analysis, data from a total of 11 panel waves are used. The dependent variable is the frequency of church attendance, which was originally measured on a 4-point scale and converted for the purpose of the regression analysis into the average annual church attendance, where 1 stands for 0 annual visits, 2 for 5, 3 for 12 and 4 for 52 annual visits. The study examines how the frequency of church attendance changes in the course of three biographical transitions (marriage, divorce and widowhood).

The research design has the following logic: Persons are observed at least once in the initial state, for example in the state unmarried. In case of a change to the target state – here: at a marriage – persons are observed for up to five calendar years in this target state. The effects shown in the regression models indicate by how much the church attendance frequency changes on average in the period after the event, compared to the period before the event.

According to model 1, the transition to the first marriage has a positive effect on the subsequent frequency of church attendance ($b = 0.75$). After marriage, the frequency of church attendance is on average 0.75 annual visits higher than in the period before. A divorce or widowhood also affect the frequency of church attendance. In the period after the divorce, the frequency of church attendance decreases by an average of 0.82 visits per year compared to the period before the divorce. After widowhood, the frequency of church service visits is significantly higher, on average by 2.27 annual visits, than before.

In addition to the event indicators, which are coded with 0 before the event and with 1 from the year of the event, all models contain control variables for age

---

[5] The RE-model belongs to the group of multilevel models with random effects mentioned above.

Table 9.2 Changes in the annual frequency of church attendance in the course of different biographical transitions. (Fixed-effects regression models, b-coefficients, t-values in parentheses). (Source: SOEP data for West Germany)

| | Model | | |
	1	2	3
Transition to first marriage	0.75***		
	(3.2)		
First divorce		−.82***	
		(−2.9)	
Widowhood			2.27***
			(3.5)
Age	0.01	0.08**	0.06
	(0.3)	(2.0)	(1.2)
Period 1996–1998 (Ref.: 92–95)	−0.78***	−0.32	−0.58**
	(−3,.5)	(−1.7)	(−2.2)
Period 1999–2003 (Ref.: 92–95)	−1.62***	−1.04***	−1.08***
	(−4.7)	(−3.6)	(−2.6)
Period 2005–2007 (Ref.: 92–95)	−2.40***	−1.35***	−1.69***
	(−4.6)	(−3.1)	(−2.7)
n (persons)	7.447	13.258	8.284
r^2 (overall)	0.01	0.03	0.01

Model 1: Persons in the initial state unmarried up to 40 years;
Model 2: Persons in the initial state married in the first marriage;
Model 3: Persons in the initial state married from 45
*$p \leq 0.10$; **$p \leq 0.05$; ***$p \leq 0.01$

and period effects. Compared to the reference period (1992–1995), the frequency of church attendance decreases across all ages with advancing calendar time. The significant age effect in model 2 means that the annual frequency of church attendance of a person increases by 0.08 units per year that they age.

Within the framework of the presented FE estimation, it is thus only examined how the frequency of church attendance changes when an event occurs (marriage, divorce, widowhood), or as a result of the increase of a continuous variable, such as age (variance *within* persons). Time-constant characteristics such as birth cohort cannot be included in the FE regression model, as here all observed or unobserved differences *between* persons that do not change are statistically controlled.

Finally, the reader is referred to selected further literature. Very understandable and conceptually oriented introductions to causal analysis with panel data are provided by Allison (1994) and Halaby (2004). More detailed presentations on panel analysis can be found in Andreß et al. (2013) or Wooldridge (2019). An advanced introduction is provided by Cameron and Trivedi (2005). The same authors have also written a book on the implementation of panel regression in Stata (Cameron and Trivedi 2010). For a compact and application-oriented overview of fixed-effects regression methods, see Allison (2009).

9.4   Covariance-Based Path and Structural Equation Models

In ordinary regression models, such as linear OLS regression, the user estimates the effect of one or more independent variables on a single dependent variable. In the context of covariance-based path analysis[6], on the other hand, several regression models are linked together. To illustrate this directly with an example, let us take a look at Fig. 9.7.

The calculated path model contains a total of three dependent variables, which are also referred to as endogenous in the terminology of path models. These are religiosity (scale of church attendance frequency and religious self-assessment), a dichotomous variable for full-time employment, and a variable consisting of five items on a traditional gender role orientation.[7] The model also contains an exclusively independent (exogenous) variable with the level of education in years.

The graphical arrangement of the variables is not given externally, for example by a statistical program, but is oriented to the theoretical considerations of the

---

[6] Unlike the covariance-based path analysis presented here, the historical predecessor, classical path analysis, is based on the principle of decomposition of regression coefficients. These two types of procedures have mathematically completely different foundations and should not be confused with each other.

[7] Examples are the items "Preschool children suffer when the mother is employed" or "Being a housewife is more fulfilling than a professional activity". The reliability of the scale is very good with a Cronbach's α of 0.81. The low case number of $n = 677$ women is explained by the fact that the items on gender role orientation were only asked of a subpopulation of the ALLBUS survey of the year 2002.

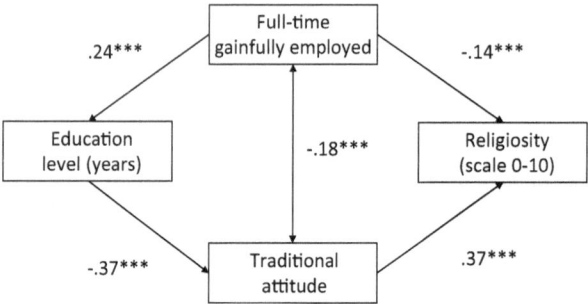

Fig. 9.7 Covariance-based path model of mediated influences of women's educational level on their religiosity. (Source: GGSS 2002; ***: $p < .001$, $n = 677$, model fit: $\chi^2 = 0.01$ (df $= 1$), $p = .093$; RMSEA: 0.00; CFI: 1.0; shown are standardized effects)

user. This illustrates an important characteristic of covariance-based path models: The starting point for testing such models is the development of a theory and the formulation of hypotheses derived from it. Path analyses thus aim at testing and not developing theories, they have a confirmatory and not an exploratory character.

In Fig. 9.7 it becomes clear that two variables, the woman's level of education and her religiosity, are not connected by an arrow. Here, the theoretical assumption is that education does not have a direct but only an indirect effect—via the employment status and the gender role orientation—on religiosity. The missing connection between the variables level of education and religiosity can be regarded as a model restriction that leads to the fact that the path model shown is no longer saturated—all variables are related to each other—but has one degree of freedom. In terms of content, the restriction made is equivalent to the assumption that the direct effect of education on religiosity is equal to zero.

How to conceptualize the principle of statistical model building is illustrated in Fig. 9.8. In his or her theory, the researcher formulates assumptions about reality, which he or she can formalize in a model. To test the theoretical model, data are collected. Whether the model fits the observed data can be assessed in covariance-based path models using various *Goodness-of-fit* measures and test statistics. With these measures, not only—as in ordinary regression analyses—is it checked whether the theoretically assumed effects, for example the relationship between gender role orientation and religiosity, are substantial. The test of model fit also takes into account whether the imposed restrictions—here the omission of a connection between education and religiosity—are appropriate, that is, compatible with the observed data. If no significant discrepancies between model and data

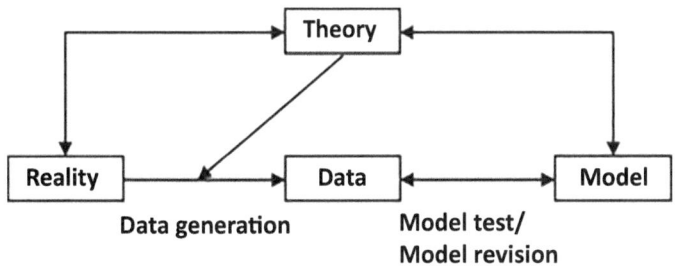

Fig. 9.8   The principle of statistical modeling. (Own representation)

are found, the found model can be statistically accepted and interpreted meaningfully. This is the case in the present example. Education 'in itself' thus does not have a direct influence on religious practice, but has—in the sense of the mediation discussed in Chap. 7—a mediated influence via the employment status and the traditional attitudes.[8]

Basically, even with a fitting model, it should be noted that other, untested model variants may show an equally good or better model fit. If the model is completely or partially rejected by the observed data, either the model can be revised or the rejection can be accepted (see Reinecke 2005, p. 9 ff.). However, extensive model modifications to improve the fit endanger the confirmatory character of the path analysis. Moreover, path models—as sometimes suggested—do not solve the fundamental causality problems that are associated with the research design—cross-sectional or longitudinal data. For example, based on the lack of longitudinal data for the model shown in Fig. 9.7, it cannot be ruled out that not the employment status affects religiosity, but the opposite direction of causality is present.

An extension of covariance-based path models are confirmatory factor analyses. Unlike exploratory factor analyses, they also rely on a well-developed theoretical measurement model, in which it is assumed that certain latent person characteristics, such as personality traits like neuroticism, determine the response

---

[8] In Chap. 7 a procedure for testing mediated (indirect) effects, the Sobel test, was already discussed. The analysis program Mplus, with which the path model calculated in Fig. 9.7 was calculated, also allows the multivariate test of indirect effects. According to this, both indirect effects of education (via full-time employment, $\beta = -0.034$ ($= 0.24 \cdot -0.14$) and via traditional attitudes, $\beta = -0.14$) are significant at the 1 % level.

to items in the questionnaire. In the measurement model, the theoretically postulated relationships between the observed and the latent variables are then estimated. The aim of confirmatory factor analysis is to test the fit between the empirical data and this theoretically based measurement model.

The development in the field of path or structural equation models has now advanced so far that a wide range of options are available to the user. For example, it is not only possible to combine confirmatory factor analyses (measurement models) with path models (structural models) in a structural equation model with latent variables. One advantage of these complex structural equation models is that the relationships between the latent variables are estimated free of measurement errors. This is an important difference from conventional regression analyses. Here, it must be implicitly assumed that the independent variables are error-free or that the manifest variables used there adequately reflect the latent constructs that often underlie them.

In addition, it is now unproblematic to estimate path or structural equation models with dichotomous or categorical endogenous (dependent) variables. Multilevel, event data or panel regression models can also be specified as path or structural equation models with modern software programs (Muthén 2002). A further advantage of path analyses is also, as shown, the inferential statistical test of indirect effects in the context of a mediation analysis (Bollen 1987).

As introductory literature in the field of path or structural equation models, Ullmann and Bentler (2012) and Schumacker and Lomax (2004) are recommended. General advantages and disadvantages of structural equation models are discussed by Nachtigall et al. (2003). Computer applications are explained for the program Mplus by Byrne (2012) and for Stata by Acock (2013).

9.5   Meta-Analyses

In many areas of the social sciences, so-called meta-analyses have established themselves as a method of research synthesis in the last 3 decades (cf. for the following explanations Klein et al. 2013). Meta-analyses are understood as a quantitative summary of published research results from different studies on the same question according to fixed rules. Compared to so-called qualitative reviews of research results, which can be found as a separate section in almost every scientific journal article, this quantification is associated with a standardization and systematization of the state of research. With a sufficiently large number of individual studies, meta-analysis can also explain the variability of research results quantitatively and draw methodologically sound conclusions. In the following, this research principle will be explained in more detail and discussed with regard to the possibilities of application in sociology.

Since the 1970s, meta-analysis has established itself as a method to examine and evaluate diverse, but often inconsistent results of empirical research—especially in medical and pharmaceutical research, but also in psychology and occasionally in pedagogy (cf. Glass 1976; Mann 1990; Hunt 1997). The starting point of the corresponding applications of meta-analysis in all these research fields is the confusion of a multitude of empirical results—not least due to the fact that the field of scientific investigations, research results and publications is growing extremely fast. However, with meta-analysis, a method is available that can solve the problems of previous research synthesis—and here especially the subjectivity of the selection and assessment of the studies—and is able to provide quantitatively statistically sound statements about the cumulative state of research and thus a well-founded statement about the state of knowledge.

The starting point of this research tradition is a work by Glass, in which he proposes meta-analysis as "analysis of analyses" (Glass 1976, p. 3)—in addition to the analysis of primary data and secondary data analysis as methods of reanalysis of existing data. While in the early years of meta-analysis the search for relevant literature on a clearly defined research problem was a major problem, today there are numerous easily accessible studies on many social science research questions. In the field of pharmaceutical and medical research, the accessibility is certainly different to evaluate. The most important basic concept of meta-analysis is the effect size. Due to different operationalizations and measurements, the individual effects of different studies are usually not directly comparable. To solve this problem, the individual effect sizes of the variables of interest are standardized: "Most important was the effect size (…): the mean difference on the outcome variable between treated and untreated subjects divided by the within group standard deviation" (Glass 1976, p. 6). In this way, it is possible to evaluate the results of smaller samples appropriately and to include them in the overall view. The first meta-analysis deals with the question of how psychotherapies influence well-being, taking into account both different outcome variables and different forms of therapy. Since 1976, the number of meta-analyses has steadily increased, especially in medicine, pharmacology, public health research, empirical education and psychology, where it is referred to as the method "for the systematic review of randomized controlled trials" (Mosteller and Colditz 1996, p. 2).

Some objections are almost routinely raised against meta-analysis, which can be summarized under three keywords: i) garbage in—garbage out, ii) oranges and apples, and iii) publication bias (Rosenthal and DiMatteo 2001, p. 66 ff.). The first argument refers to the problem that the methodological quality of the studies included in a meta-analysis can vary greatly. If studies differ significantly in the quality of measurements or design, they should not be considered together. It was therefore suggested early on (cf. Rosenthal and DiMatteo 2001, p. 67) to capture the quality of the studies and to include it as a control variable in the meta-analysis. The second argument aims at the incomparability of the operationalizations and measurements used. Here, too, these differences can be taken into account as moderator variables in a meta-analysis, and moreover: "It is a good thing to mix apples and oranges, particularly if one wants to generalize about fruits" (Rosenthal and DiMatteo 2001, p. 68). A third point of criticism relates to the publication strategies of authors, but also of journals (Hunt 1997, p. 118 ff.). Statistically significant results have an increased chance of being submitted and published. The influence of the relevant factors is overestimated if studies that have not produced a significant result are published less frequently. The answer to this so-called 'file-drawer problem' lies in the calculation of the number of non-significant studies that would be necessary to change the results of a meta-analysis in terms of content. Usually, this number is in a range that makes it unlikely that the content interpretations would have to be changed.

Even though the three difficulties mentioned address important problems, they do not pose any fundamental obstacles to meta-analyses as a method of research synthesis. On the contrary: A large number of analyses, especially in the so-called 'life sciences', show the enormous advantage of meta-analyses over methods such as qualitative research reviews or the so-called vote-counting, the mere counting and weighing of positive and negative research findings. Meta-analyses offer the possibility to make quantitatively statistically sound statements about the state of research in various fields and thus to provide real contributions to the advancement of knowledge, which can save costs and lives in medicine, for example.

Now the consequences of sociological research are rarely so dramatic, or their drama is at least rarely so immediately visible. Nevertheless, one can ask oneself why these methods are only rarely used in sociology? One of the reasons for the wide spread of meta-analyses in the above-mentioned sciences is the large number of rather small studies on the same question. The investigation of sociological questions, on the other hand, often requires large case numbers. Accordingly, the relevant research focuses on a few large-scale studies, with which the infrastructure of empirical social research has improved considerably in recent decades. The low prevalence of meta-analyses in sociology, however, is not only explained by the rather small number of a few large-scale studies in many research fields, but there are also research-logical reasons that suggest or perhaps even force doubts about the applicability of meta-analysis in sociology. These concern the calculation of effect sizes that came about in very different ways and are not based on an experimental design that is characterized by the constancy of a large number of factors. The background of the problem is that a factor of influence under investigation is often closely related to other factors (third variables) and ultimately almost always an ex-post-facto design can be found. The analyses of interest then usually work with moderation and mediation analyses (cf. chap. 7).[9] In contrast to sociology, empirical research in particular in medicine and psychology is strongly influenced by experiments, and a random allocation of the subjects to the experimental and the control group (the so-called randomization) ensures that third variables do not create a systematic difference between the two groups, so that differences of the variable under investigation can be attributed exclusively to the experimental variable. The results found in randomized experiments can therefore be relatively easily calculated in a meta-analysis.[10]

Against the background of these problems, which meta-analysis is confronted with especially in sociology, one does not have to stop trying to achieve a research synthesis. It is quite possible to summarize and compare different studies not on the basis of published results, but on the basis of the original data. This form of meta-analysis refers to the units of analysis contained in the original data instead of regression coefficients, in which published results often crystallize. The term meta-analysis refers not to the units of analysis—original data versus regression coefficients—but to the overarching summary and comparison of different studies on the basis of the original units of analysis, as they underlie the primary and secondary analyses. Also in the literature on meta-analysis, a corresponding procedure has been discussed in recent years: "When the meta-analyst has access to all of the original data from each study, the meta-analysis may be referred to as

---

[9] This is a fundamental problem of meta-analyses in sociology, which is also seen in the corresponding literature on meta-analysis: "Coefficients from bivariate and multivariate methods differ according to their magnitude and standard errors. (...) meta-analysis misses adequate procedures of multivariate result integration" (Wagner and Weiß 2006, p. 488, translated by authors). "Multivariate relationships present special challenges to meta-analysis. (...) the varying sets of independent variables across regression equations complicates their synthesis" (Lipsey and Wilson 2001, p. 67). "Especially the integration of partial coefficients is a problem that has not yet been satisfactorily solved" (Wagner and Weiß 2003, p. 35, translated by authors).

[10] Of course, experiments also differ in their quality, for example with regard to the quality of the blinding of the study or similar factors. However, there is then the possibility to include these quality criteria in a meta-analysis and to interpret the results under control of the methodological quality of the study (cf. Borestein et al. 2009).

an individual participant data (or individual patient data) meta-analysis" (Borenstein et al. 2009, p. 316; cf. also Sutton et al. 2008; Berlin et al. 2002; Sutton and Higgens 2008). For further reading, we recommend Borenstein et al. (2009).

9.6   Afterword

At the end of this chapter, the readers should have developed a first idea of what kinds of questions the respective method is suitable for and what evaluation possibilities it conceptually offers. The learning objective is thus rather modestly formulated: It is a first introduction to the respective field. In order to be able to apply the respective method in practice, a deeper engagement with the respective special literature is indispensable. Beyond the theoretical reading, it is always advisable to practice the respective method in practice, that is, with empirical data and the relevant data analysis programs.

References

Acock, Alan C. 2013. *Discovering structural equation modeling using Stata.* College Station: Stata Press.

Allison, Paul D. 1994. Using panel data to estimate the effects of events. *Sociological Methods & Research* 23:174–199. https://doi.org/10.1177/0049124194023002002.

Allison, Paul D. 2009. *Fixed effects regression models. Quantitative applications in the social sciences 160.* Thousand Oaks: Sage. https://doi.org/10.4135/9781412993869.

Allison, Paul D. 2014. *Event history and survival analysis. Regression for longitudinal event data.* Quantitative applications in the social sciences, Vol. 46. Thousand Oaks: Sage.

Andreß, Hans-Jürgen., Katrin Golsch, und Alexander W. Schmidt. 2013. *Applied panel data analysis for economic and social surveys.* Wiesbaden: Sringer. https://doi.org/10.1007/978-3-642-32914-2.

Berlin, Jesse A., Jill Santanna, Christopher H. Schmid, Lynda A. Szczech, und Harold I. Fledman. 2002. Individual patient- versus group-level data meta-regressions for the investigation of treatment effect modifiers: Ecological bias rears its ugly head. *Statistics in Medicine* 21:371–387. https://doi.org/10.1002/sim.1023.

Blossfeld, Hans P. 2010. Survival- und Ereignisanalyse. In *Handbuch der sozialwissenschaftlichen Datenanalyse*, Eds. Christof Wolf und Henning Best, 995–1016. Wiesbaden: VS Verlag. https://doi.org/10.1007/978-3-531-92038-2_37.

Blossfeld, Hans P., Götz. Rohwer, und Thorsten Schneider. 2019. *Event history analysis with Stata.* New York: Routledge.

Bollen, Kenneth A. 1987. Total, direct, and indirect effects in structural equation models. In *Sociological methodology*, Eds. Clifford C. Clogg, 37–69. Washington, D.C.: American Sociological Association. https://doi.org/10.2307/271028.

Borenstein, Michael, Larry V. Hedges, Julian P. T. Higgens, und Hannah R. Rothstein. 2009. *Introduction to meta-analysis*. Chicester: Wiley. https://doi.org/10.1002/9780470743386.

Broström, Göran. 2012. *Event history analysis with R*. Boca Raton: CRC Press. https://doi.org/10.1201/9781315373942.

Brüderl, Josef, und Andreas Diekmann. 1995. The log-logistic rate model: Two generalizations with an application to demographic data. *Sociological Methods and Research* 24:158–186. https://doi.org/10.1177/0049124195024002002.

Byrne, Barbara M. 2012. *Structural equation modeling with Mplus. Basic concepts, applications, and programming*. New York: Routledge.

Cameron, A. Colin., und Pravin K. Trivedi. 2005. *Microeconometrics: Methods and applications*. Cambridge: Cambridge Univerity Press. https://doi.org/10.1017/cbo9780511811241.

Cameron, A. Colin., und Pravin K. Trivedi. 2010. *Microeconometrics using stata*. College Station: Stata Press.

Cleves, Mario, Roberto G. Gutierrez, William Gould, und Yulia A. Marchenko. 2010. *An introduction to survival analysis using Stata*. College Station: Stata Press. https://doi.org/10.18637/jss.v012.b01.

Garson, David G. 2019. *Multilevel modeling. Applications in Stata, IBM SPSS, SAS, R und HLM*. Thousand Oaks: Sage.

Glass, Gene V. 1976. Primary, secondary, and meta-analysis of research. *Educational Researcher* 5:3–8. https://doi.org/10.3102/0013189x005010003.

Goldstein, Harvey. (2010). Multilevel statistical models. Chichester: Wiley. https://doi.org/10.1002/9780470973394.

Guo, Guang. 1993. Event-history analysis for left-truncated data. *Sociological Methodology* 23:217–243. https://doi.org/10.2307/271011.

Guo, Guang, und Hongxin Zhao. 2000. Multilevel modeling for binary data. *Annual Review of Sociology* 26:441–462. https://doi.org/10.1146/annurev.soc.26.1.441.

Halaby, Charles N. 2004. Panel models in sociological research. *Annual Review of Sociology* 30:507–544. https://doi.org/10.1146/annurev.soc.30.012703.110629.

Hank, Karsten, Michaela Kreyenfeld, und Katharina Spieß. 2004. Kinderbetreuung und Fertilität in Deutschland. *Zeitschrift für Soziologie* 33:228–244. https://doi.org/10.1515/zfsoz-2004-0303.

Hox, Joop. 2010. *Techniques and applications*. New York: Routledge. https://doi.org/10.4324/9780203852279.

Hunt, Morton. 1997. How science takes stock. *The story of meta-analysis*. New York: Sage. https://doi.org/10.1136/bmj.317.7165.1088b.

Klein, Thomas, Johannes Kopp, und Ingmar Rapp. 2013. Metanalysen mit Originaldaten. Ein Vorschlag zur Forschungssynthese in der Soziologie. *Zeitschrift für Soziologie* 42:221–236. https://doi.org/10.1515/zfsoz-2013-0304.

Lipsey, Mark W., und David B. Wilson. 2001. *Practical meta-analysis*. Thousand Oaks: Sage.

Lois, Daniel. 2011. Wie verändert sich die Religiosität im Lebensverlauf? Eine Panelana-lyse unter Berücksichtigung von Ost-West-Unterschieden. *Kölner Zeitschrift für Soziologie und Sozialpsychologie* 63: 83–110. https://doi.org/10.1007/s11577-010-0124-z.

Luke, Douglas A. 2019. *Multilevel modeling. Quantitative applications in the social sciences 143*. Thousand Oaks: Sage. https://doi.org/10.4135/9781412985147.

Mann, Charles M. 1990. Meta-analysis in the breech. *Science* 249:476–480. https://doi.org/10.1126/science.2382129.

Mills, Melinda. 2011. *Introducing survival and event history analysis*. Thousand Oaks: Sage. https://doi.org/10.1177/026850911427995.

Mosteller, Frederick, und Grahamn A. Colditz. 1996. Understanding research synthesis (meta-analysis). *Annual Review of Public Health* 17:1–23. https://doi.org/10.1146/annurev.pu.17.050196.000245.

Muthén, Bengt O. 2002. Beyond SEM: General latent variable modeling. *Behaviormetrika* 29:81–117. https://doi.org/10.2333/bhmk.29.81.

Nachtigall, Christof, Ulf Kroehne, Friedrich Funke, und Rolf Steyer. 2003. (Why) should we use SEM? Pros and cons of structural equation modeling. *Methods of Psychological Research Online* 8:1–22.

Rabe-Hesketh, Sophia, und Anders Skrondal. 2022. *Multilevel and longitudinal modeling using Stata. Vol. I: Continuous response*. College Station: Stata Press.

Reinecke, Jost. 2005. *Strukturgleichungsmodelle in den Sozialwissenschaften*. München: Oldenbourg.

Rosenthal, Robert, und M. Robin DiMatteo. 2001. Meta-analysis: Recent developments in quantitative methods for literature reviews. *Annual Review of Psychology* 52:59–82. https://doi.org/10.1146/annurev.psych.52.1.59.

Schulze, Alexander, Felix Wolter, und Rainer Unger. 2009. Bildungschancen von Grund-schülern: Die Bedeutung des Klassen- und Schulkontextes am Übergang auf die Sekundarstufe I. *Kölner Zeitschrift für Soziologie und Sozialpsychologie* 61:411–435. https://doi.org/10.1007/s11577-009-0072-7.

Schumacker, Randall E. and Richard G. Lomax. 2004. *A beginner's guide to structural equation modeling*. New York: Psychology Press. https://doi.org/10.4324/9781410610904.

Singer, Judith D., und John B. Willett. 2003. *Applied longitudinal data analysis. Modeling change and event occurrence*. Oxford: University Press. https://doi.org/10.1093/acprof:oso/9780195152968.001.0001.

Snijders, Tom, und Roel Bosker. 2011. *Multilevel analysis. An introduction to basic and advanced multilevel modeling*. London: Sage.

Sutton, A.J., D. Kendrick, und C.A.C. Coupland. 2008. Meta-analysis of individual- and aggregate-level data. *Statistics in Medicine* 27:651–669. https://doi.org/10.1002/sim.2916.

Sutton, Alexander, und Julain P. T. Higgens. 2008. Recent developments in meta-analysis. *Statistics in Medicine* 27:625–650. https://doi.org/10.1002/sim.2934.

Ullmann, Jodie B., and Peter M. Bentler. 2012. Structural equation modeling. In *Handbook of Psychology*, Ed. B. Weiner, 661–690. Wiley. https://doi.org/10.1002/9781118133880.hop202023

Wagner, Michael, und Bernd Weiß. 2003. Bilanz der Scheidungsforschung. Versuch einer Meta-Analyse. *Zeitschrift für Soziologie* 33:29–49. https://doi.org/10.1515/zfsoz-2003-0102.

Wagner, Michael, und Bernd Weiß. 2006. On the variation of divorce risk in Europe: Findings from a meta-analysis of European longitudinal studies. *European Sociological Review* 22:483–500. https://doi.org/10.1093/esr/jcl014.

Wooldridge, Jeffrey M. 2019. *Introductory econometrics: A modern approach*. Nashville: Southwestern Publishing House.